Six Simple Twists

T0251485

The Pleat Pattern Approach
to Origami Tessellation Design

Second Edition

Six Simple Twists

The Pleat Pattern Approach
to Origami Tessellation Design

Second Edition

Benjamin DiLeonardo-Parker

CRC Press
Taylor & Francis Group
Boca Raton London New York

CRC Press is an imprint of the
Taylor & Francis Group, an **informa** business

AN A K PETERS BOOK

First edition published in 2016

by CRC Press
6000 Broken Sound Parkway NW, Suite 300, Boca Raton, FL 33487-2742
and by CRC Press
2 Park Square, Milton Park, Abingdon, Oxon, OX14 4RN

© 2021 Taylor & Francis Group, LLC
CRC Press is an imprint of Taylor & Francis Group, LLC

Library of Congress Control Number: 2020938234

ISBN: 978-1-138-31188-6 (pbk)
ISBN: 978-1-138-31192-3 (hbk)
ISBN: 978-0-429-45856-9 (ebk)

Typeset in Avenir LT Std
by Nova Techset Private Limited, Bengaluru & Chennai, India

Visit the Taylor & Francis Web site at
http://www.taylorandfrancis.com

and the CRC Press Web site at
http://www.crcpress.com

Contents

Contents

Preface

This text explores origami pleat patterns. Pleats are overlaps of paper. When multiple pleats converge on each other, they displace an area of that paper. The area displaced depends on factors such as the number of pleats, their width, and their angle of intersection. Multiple pleat intersections can be connected to create a network of interacting parts, and the visual qualities are often stunning. These patterns are commonly called *origami tessellations*. When I started attending origami conventions in 2007, there weren't many folders of these tessellations (or tessellators, as we call ourselves). The problem was the difficulty in teaching the techniques. Traditional origami is generally taught in diagrams that give step-by-step instructions.

With tessellations, there are several simultaneously moving parts of the paper, creating a more complex procedure. Teachers and authors would, at one extreme, instruct a specific design without digging into the potential for variation, or at the other extreme they would explore generalized mathematics that was too advanced for many folders. I have endeavored to find a middle ground: a balance where the reader can learn to fold tessellations and use that learning to design their own works afterward. This book features my methods of design, but there are other schools of thought. I encourage readers to find their own styles and techniques; explore, fold, and have fun!

Though this book focuses on origami tessellation design, there are many skills required before I begin. I will spend time on the individual components of tessellations before branching into combining those components. You can use the Table of Contents to skip ahead to the parts you'd most like to explore.

This book is divided into three chapters.

- Chapter 1 is an overview of the *six simple twists*—curated forms that will introduce a new folder to some basic twists.
- Chapter 2 involves techniques for combining those twists into full pleat patterns.
- Chapter 3 contains alternative twists and an in-depth analysis of twist creation.
- Chapter 4 includes some final thoughts about this book, and a guest section from a colleague, Matt Benet, on further mathematical analysis.

If you have no experience creating origami tessellations, I recommend starting with Chapter 1. This will instruct you on how to fold a hexagon, a triangle grid, and teach you the six simple twists. If you have some experience, or if you've folded an origami tessellation and want additional skills, you could begin with Chapter 2. If you are experienced and would like to explore twist designs, pushing the limits of your understanding, you might enjoy starting with Chapter 3; at this point, the content will delve more heavily into mathematics, and continue that through to Chapter 4.

Acknowledgments

This book started as a solo venture to get my thoughts down and organized. After the first edition came out, I realized this was the wrong approach. With this edition, I wanted to have more collaboration, and so asked several artists to contribute work to it. I am very pleased with the work offered to the gallery by extremely talented practitioners of their art: Alessandro Beber, Joel Cooper, Ilan Garibi, Melina "Yureiko" Hermsen, Michał Kosmulski, Robert Lang, Christine Dalenta, Halina Rościszewska-Narloch, Robin Scoltz, and Helena Varrill.

As the project grew, I reached out for assistance in proofreading and was amazed at the response I got. I wish to thank all of my proofreaders, as well as those who contributed to editing or taking photographs and testing diagrams, especially Andrew Albin, Matt Benet, Alicia Gilbride, Michał Kosmulski, Uyen Nguyen, Mimansa Vahia, and Kerstin Wienl. You've all helped me to describe concepts that are particularly complicated in a manageable and more descriptive manner, and your influence comes through in the final design.

While I was working on the initial parts for the second edition, I began collaborating with Andrew Fisher and Matt Benet to develop a notation system for pleat intersections. Through this system and later analyses, I was able to deepen my own understanding of tessellations tremendously; I had no choice but to delay the finishing of the book so that we would have time to explore more fully. I do not regret that delay at all, since I believe it has elevated the text far beyond what it would have been otherwise. I look forward to continuing the studies with you!

About the Author

Born in Pittsburgh and living in Connecticut, Benjamin DiLeonardo-Parker has been an active student of origami tessellations since 2007. He has taught and exhibited at origami conventions and art shows internationally, including Chi Mei Museum (Tainan City, Taiwan), La Escuela-Museo Origami de Zaragosa (Zaragosa, Spain), The Science Museum Oklahoma (Oklahoma City, OK), The Museum of Mathematics (New York, NY), The Japan Information and Culture Center (Washington DC), The New Britain Museum of American Art (New Britain, CT), and The Cooper Union Gallery (New York, NY).

Outside of art, Ben teaches high school mathematics to students with uncommon learning styles and incorporates origami into his classes as often as he can. Ben approaches his artwork from a holistic standpoint, preferring to view origami as an entry into the vast network of disciplines to which it is connected. This has led him to extend his knowledge of education, engineering, mathematics, CNC fabrication, paper arts, fashion, alternative photography, and other studies. When not teaching high school math, Ben operates a workshop in Essex, CT, out of which he creates artwork and runs classes on origami design.

He views his practice of origami as cyclical and recursive. Origami is connected to such a vast network of disciplines, each with its own siren's call. Each flavor, each culture, cycles back onto its own basics over and over, swirling and interacting with previous knowledge, each enhancing the others in some way.

Edition-al Material

This second edition of *Six Simple Twists* is rather different from the first. The first edition offered foundational techniques for creating new origami tessellations, and while it made strides in that direction, I felt that there was much more to discuss. In this edition, almost every aspect of the book has been modified and improved, including the quality and size of the photos, the order in which concepts are presented, and the number of demonstrated patterns. Finally, to further explore the underlying mathematics of the craft, I have added a new, in-depth analysis of what makes a twist actually twist and fold flat in Chapter 3.

In addition, the reader will find:

- Several works by and references to other contemporary origami artists and researchers
- More basic skill instruction than the first edition, including but not limited to alternate methods for creating a twist, to help newer folders ease into more advanced topics
- Methods for manipulating the elements of an origami tessellation–pleats and their intersections–to provide even more tools for creation
- A simplified, yet more robust system for pleat intersection notation

Chapter 1

Why Study Pleat Patterns?

Although paper is an often underestimated artistic medium, there are several compelling features in its favor. As a material, it can vary from ordered and crisp to chaotic and fibrous, depending on how it is manufactured and manipulated. Within this range, the sheer amount of expression made possible by a sheet of paper is massive.

Why study origami? Other paper art forms exist, such as collage, paper sculpting, and paper engineering. Why confine artistic expression to a single uncut sheet of paper, following traditional origami approaches? Personally, I enjoy just grabbing a sheet and a table and going at it. No mess, nothing to clean up, and no scissors to carry around. If I can fold something out of one sheet of paper, why should I use two? If I can fold without any cuts, why cut at all? It's a challenge, one on which origami artists thrive.

Why study origami tessellations, a substyle of origami that's even more of a niche? In many ways, that's one of its biggest selling points. The study of origami tessellations is unique, with a small, supportive international community of designers. The terms of the artform are still being written, the techniques are still being explored, and the horizon is still expanding. If you have designed a pattern of sufficient complexity, you can say with reasonable certainty that you're the only person to have folded it—though it should be noted that I prefer to use the term "independent discovery" when speaking about origami tessellations rather than "design." This creation feeling is incredibly powerful and entices you to continue discovering new patterns. It's the study of a kind of poetry in the language of geometry. And like many languages, once you've mastered one dialect, mastery of others comes more easily. I know several origami artists who came from or went on to master complex knitting or sewing designs, for example. The knowledge you build by studying pleat patterns translates to other activities and artforms.

Sometimes, the reason for folding is simply to discover what a concept looks like on paper. You're thinking about what's possible to fold and you stumble upon a twist that looks really interesting. You start to wonder what the twist will look like tiled, and suddenly you find yourself halfway through a preparatory grid before you remember how you got there. It's a compulsion, the best of its kind. I hope this book helps you to appreciate this compulsion as well.

1.1 Basics and Preparation

For beginners, I recommend folding from copy paper, since it's relatively strong, does not crack, and is generally plentiful and cheap. You can draw on it and mess it up, and this is critically important to the learning process. As you go through the exercises and your confidence improves, you may consider coloring the paper or using higher quality material.

In the photos throughout this book, I use a paper brand called Elefantenhaut—or elephant hide—a German paper used for bookbinding. This has become a favorite of the geometric origami community due to its strength and the crisp folds that come from it. I have also seen origami artists use the brand Stardream for similar reasons. Wax paper folds well and allows you to see multiple layers of paper at once, which can look quite beautiful. Certain types of handmade paper can work if they are made from long, strong fibers. I've typically seen abaca paper work well for this style of folding. Whatever you use should be strong enough to handle many creases but made in such a way that it won't crack from a crisp crease. Different papers may take more or less abuse as you're folding, and only experience will tell you how to treat a particular paper.

When coloring the paper, since there is so much folding and unfolding, try to use thinner substances, such as watercolors or pastels. Thicker paints such as acrylics will likely crack under the stress of movement. I encourage you to try many different types of papers (or even other foldable materials) and colorations to find your own style!

1.2 How Pleat Patterns Differ from Traditional Origami

In traditional folding, there is often a sequenced series of steps called a diagram, and after each step, the paper reaches a stable, generally flat form. So long as the diagram is clear enough, this makes it relatively easy to understand how the paper is supposed to be folded. Typically, the fewer steps a diagram has, the more the author assumes the reader knows about the steps leading to the finished form. Diagrams become more complicated when there are multiple facets moving simultaneously, as is often the case with twist designs. We will see this complication frequently in the instructions of this book, and it can be difficult to visualize for novice folders.

Even after performing a twist multiple times, it can be surprising when it all comes together and frustrating for newer folders when it does not. To aid folders with getting the creases aligned, it is helpful to start with a *grid*, which will act as a guide for the *pleats*. This is done as a preparatory step to folding a full pleat pattern and is generally not needed for traditional origami.

A further way traditional origami and origami tessellations differ is that tessellations require a lot more unfolding than traditional models. When you prepare a grid from (for example) a hexagon, the result is the same hexagonal shape, unchanged except for the new creases. A precreased grid takes quite a bit of time to create, and this can be off-putting to some. Sometimes you will also have to unfold or partially unfold a twist you've already made to begin folding the next one. This gets easier with practice, and the confidence that you can refold anything that you've unfolded increases as you do it more. Remember that paper has a memory and to some degree will naturally want to refold into a form it has already taken.

One final major difference between these two styles of origami is the end goal. Traditional origami generally attempts to produce finished models that represent objects, such as creatures found in nature. Pleat patterns are fully capable

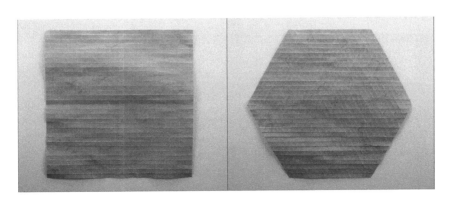

Figure 1.2.1 LEFT: Square grid. RIGHT: Triangle grid.

of representation (see Halina Rościszewska-Narloch's work in the gallery section of Chapter 3), but most of the foundational exercises are purely geometric. However, should you wish to use a pleat pattern to create something representational, it will likely have a unique and beautiful look all to itself.

1.3 Fold Parity and How to Read the Diagrams

If you have experience with origami, you will be familiar with common methods for documenting folds. The most common is the diagram, which shows several consecutive steps in a row. With origami tessellations, the folder has to maneuver multiple regions of the paper at the same time, and many "partial steps" need to be addressed in the diagrams. For this reason, I have divided each "step" into clusters of in-process maneuvers that I feel will explain better than just before-and-after photos. While reading, pay attention to my hand position in the photos. Where am I pinching? Do I have to push up from the backside of the paper? How do I keep count of spacing between features?

Every fold exists in one of two states, either mountain or valley. You could also say that there is an "unfolded" state, but we'll keep the delineation to mountain and valley for now. Whether a fold is in a mountain or valley state is called that fold's *parity*. A fold with mountain parity on one side of the paper has a valley parity when the paper is turned over, and vice versa.

Many origami practitioners are familiar with the concept of a *crease pattern*. A crease pattern (hereafter written with the conventional shorthand "CP") is what you get when you fold a work of origami and then unfold it. It does not offer any instruction as to the order of the folds, but experienced folders know that longer creases tend to be folded before the others and that chains of similar-parity folds tend to be reflections, which allows for some deduction on the part of the person reproducing a form. Figure 1.3.1 shows a traditional paper crane with the unfolded form. The right photo shows its CP.

In the diagrams and CPs, I use blue solid lines to represent mountain folds and red dashed lines to represent valley folds. For unfolded creases, I use light, solid gray or black lines, and for folds that have some bias one direction or the other, I use thin blue or red solid lines. For CPs and other figures, I use the white side of the paper to show the front and a gray fill for the reverse, as shown in Figure 1.3.2.

Figure 1.3.1 LEFT: Traditional crane. MIDDLE: Unfolded version of the crane. RIGHT: Paper crane CP.

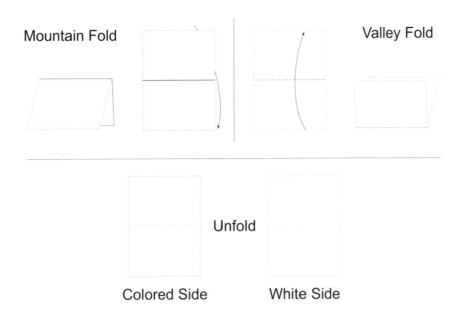

Mountain Fold

Valley Fold

Unfold

Colored Side

White Side

Figure 1.3.2 Mountain/valley folds and unfold notation.

Front and Reverse of the Paper

Most of the photographed steps in this book do not involve flipping the paper. For those that do, I outline the reverse of the paper with a dotted red border. In this case, the "front" side is the side you are viewing the most in the diagram, and the reverse is the other side of the paper. This border differentiation is used heavily in the gridding diagrams in Sections 1.6 and 2.1, as well as Section 2.35, the iso-area twist tessellation. Naturally, this indication only applies to photos that capture an entire view of the paper, on one side or the other. For diagrams that do not include photos outlined this way, assume you are still viewing the front of the piece. Always follow the directions closely and pay special attention to the word "flip" in any step.

Images here are of the front-side of the work.

Images here are of the reverse-side of the work.

Figure 1.3.3 LEFT: Front of work border. RIGHT: Reverse of work border.

Rotations

We will be dealing a great deal with elements that twist in one rotational direction or the other. As shorthand, I will replace the word "clockwise" with CW and "counterclockwise" with CCW. While most forms will rotate CCW for consistency, it should be noted that that any form with a rotation can also be created with the opposite rotation.

1.4 Folding Uniform Parallel Creases

To fold a grid, we have to create parallel creases and be reasonably certain of their accuracy.

To fold a parallel pleat, take a rectangular sheet of paper and fold some arbitrary distance up the paper. You can use the sides of the paper as guides to make sure the crease is folded perpendicularly to the edges as shown. Then unfold to reveal crease a.

To make the second crease parallel to the first, flip the paper. Flipping it to the other side and laying it on a table allows you to pinch the crease that now rises as an unfolded mountain, which is easier than pinching the former valley crease. Pinch the crease and bend it away from you, then press it flat on the table to form an overlap, as shown in Figure 1.4.2. Use the edges of the paper as guides again to keep everything aligned, then unfold. Now you can say with confidence that creases a and b are as accurately parallel as possible.

When folding pleats, we want the distance between parallel creases to be consistent across the paper. Origami lends itself naturally to multiplying or dividing by two, which equates to

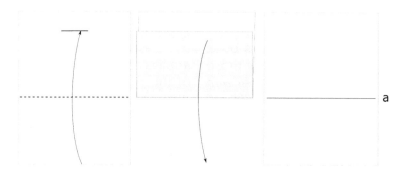

Figure 1.4.1 First crease.

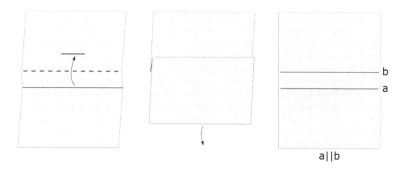

Figure 1.4.2 Second crease.

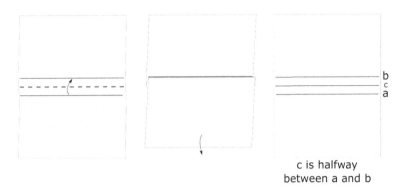

c is halfway
between a and b

Figure 1.4.3 Find the halfway crease.

doubling or halving a margin or angle, simply by folding creases and/or edges to meet one another.

To prepare the paper for pleat patterns, you will have to create a grid. The process involves subdividing the margins between creases iteratively until you've reached the desired density. You can subdivide the distance between the two creases by folding one crease to the other and creating a third parallel crease at their midpoint. Lift the mountain fold of the pleat in the same manner as described previously and fold it to meet the pleat's other crease. This will create a crease that is exactly halfway between the other two creases. This procedure is shown in Figure 1.4.3.

You can also propagate an interval between two creases throughout the paper. To do this, we take advantage of a physical property of the paper: its thickness. Beginning with a folded pleat, place your thumb on the overhanging paper, run it across the paper along the ridge felt underneath, where the pleat edge lies. Depending on the paper's thickness, this can create a mark which mirrors the valley crease lying underneath. With practice, you can create rather accurate pleats this way, and it's certainly accurate enough for the complexity of the works in this book. The process for propagation is shown in Figure 1.4.4.

As line drawings can only show so much, it is helpful to see photos of the physical techniques for folding straight creases. Here is my approach to realizing the previously shown figures with paper.

When creating the fold, I've found the most accurate method is to always fold away from yourself. Use your fingers to set the corners at the far edge of the paper as shown in the third

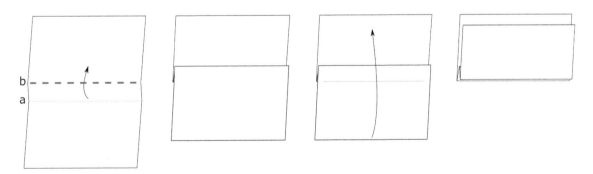

Figure 1.4.4 Duplicate an interval.

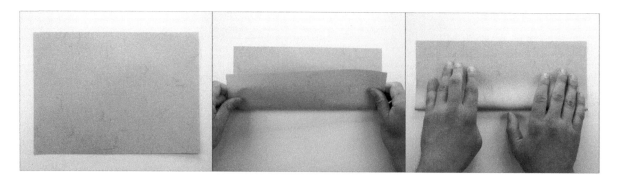

Figure 1.4.5 LEFT: Rectangular paper. MIDDLE: Valley fold in half lengthwise. RIGHT: Line up the corners.

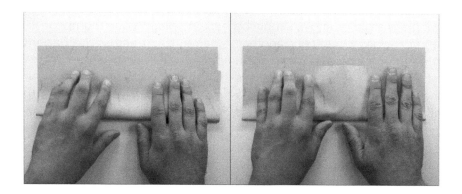

Figure 1.4.6 Watch out for kinks in the paper.

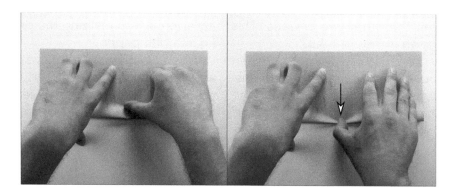

Figure 1.4.7 Draw your thumb directly back and pinch.

photo of Figure 1.4.5. Do not crease down until you have set the paper where you want it.

Once you have the paper in the position you want, drag a thumb straight toward yourself in the middle of the crease.

Then, use your open hand to start the fold. This is just the first pass, so press down only enough for the paper to bend where you are making the fold. This will condition the crease in preparation for the next pass.

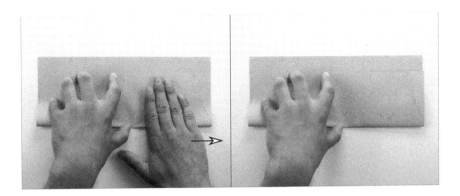

Figure 1.4.8 Condition the crease.

Then, take your thumbnail and go over that same half of the fold again, keeping constant pressure. This should result in a good, crisp fold. In Figure 1.4.9, the bottom right photo demonstrates the difference between the left and right halves of the crease, the latter having been crisply finished.

After you've used your thumb to crease both halves of the line, unfold and view the straight crease. Test yourself using the techniques described: Can you propagate a pleated interval across the rest of the paper to create a set of evenly-spaced parallel creases?

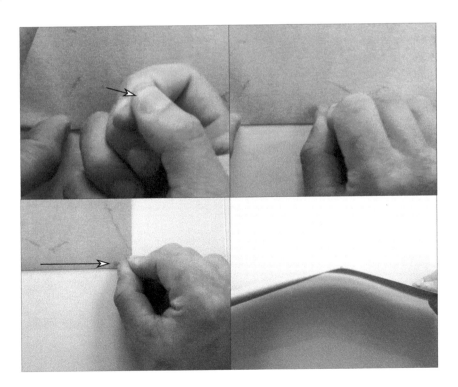

Figure 1.4.9 Finish the crease using your thumbnail.

Figure 1.4.10 Unfold.

The previously outlined steps explain how to create parallel creases as long as there are guides already in place (the edges of the paper), but what if there are no guides to help you? To explore this, start with a sheet of paper with an arbitrarily angled crease as shown in Figure 1.4.11.

How do you create a crease parallel to this one? Well, you can create a crease perpendicular to it by folding the crease onto itself as shown in Figure 1.4.12.

You do not have to crease it sharply; you are just looking for a reference to build from.

Once you have a crease perpendicular to the first, you can use the same method to create a crease perpendicular to the second and parallel to the first, as in Figure 1.4.13. Now you have three creases on the paper, the first and third being perpendicular to the second. But if they are perpendicular to the second crease, they must be parallel to each other.

And thus, you can use your distance-propagation technique and subdivision technique taught earlier.

As you're folding you may notice a few things. The first is that the paper starts to curl in a certain direction as you place a series of valley folds. In the final drawing of Figure 1.4.14, there are more valley than mountain folds on the side of the paper facing the viewer, so the paper will curl toward the viewer. To understand this, you need to think about crease bias and figure out a systematic way to neutralize the bias through a process called *backcreasing*.

Take a sheet of paper and fold a crease from edge to edge (the specifics do not matter). Then

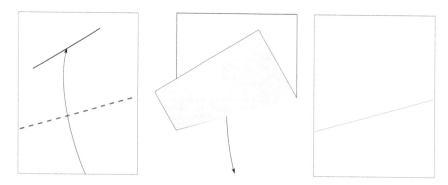

Figure 1.4.11 First crease.

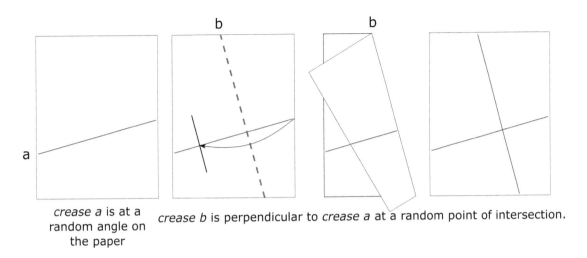

crease *a* is at a random angle on the paper

crease *b* is perpendicular to *crease a* at a random point of intersection.

Figure 1.4.12 Fold a perpendicular crease.

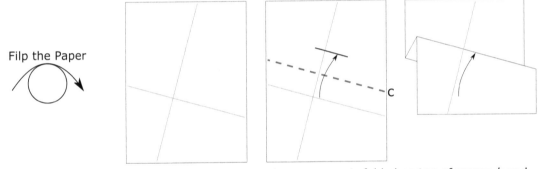

Filp the Paper

when crease *a* is folded on top of crease *b* and *crease b* stays aligned to itself, it forms valley fold *crease c*, which is parallel to *crease a*.

Figure 1.4.13 Second crease.

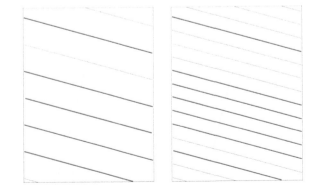

Figure 1.4.14 Propagate the interval.

take your hands off and look at the curve that the profile of the fold creates. Depending on several factors such as the type of material used, how hard you creased, and how much unfolded area lies on either side of the crease, the curve in the unfolded paper will vary greatly. What's important to note is its existence, and this should be noted as well during gridding. As parallel creases become more and more dense, they can easily start to curl and become unwieldy.

Figure 1.4.16 shows a few possible results from various methods of precreasing: This

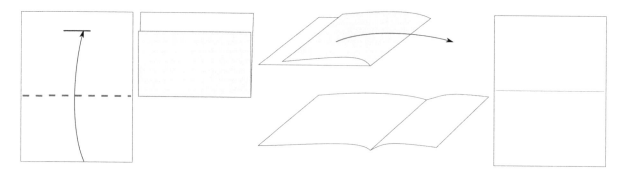

Figure 1.4.15 Crease bias and memory.

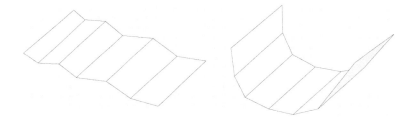

Figure 1.4.16 Curl due to crease bias.

curling effect is due to the paper *memory*. Memory, which varies between materials, refers to the tendency of the material to curl or bend around a mountain or valley fold and retain this curled or bent state after you remove your hands. You can decrease the memory of the crease—and so increase the ease of its manipulation—by *backcreasing*.

To backcrease, I like to start on the mountain fold side of a crease and lift the raw edge of the paper up and over. Then, I press until the crease begins to break in the opposite direction,

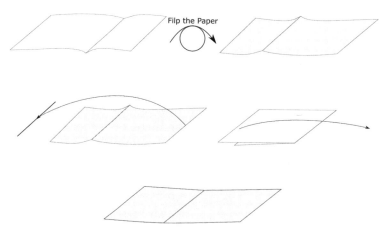

Figure 1.4.17 Backcreasing to decrease bias.

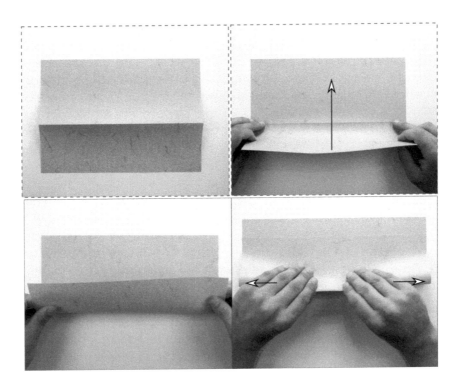

Figure 1.4.18 TOP LEFT: Flip the paper. THE REST: Reverse the crease made in Figure 1.4.10 and pinch.

changing the fold parity. Be very careful not to accidentally introduce a new crease near the one you're trying to backcrease. Note that I'm not flipping the paper but changing the crease from a mountain to a valley fold on the same side. When you unfold, the bias of the former mountain fold will be less, and the crease will have inverted to form a weak valley fold where the stronger mountain fold used to be.

Once a fold is backcreased, it still has some memory, but the crease is far more neutral and malleable than creases that are only folded

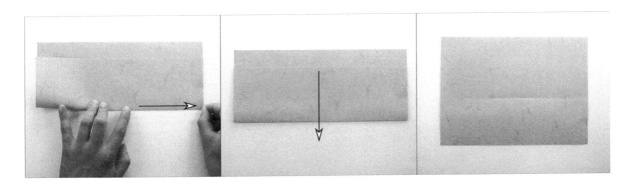

Figure 1.4.19 Crease and unfold.

once. If you wished, you could fold the crease back and forth more than once, each time switching between a mountain and valley, but also decreasing the paper's memory each time. For the sake of time, I usually stop at two iterations, an initial fold and a single backcrease.

1.5 Grid Axes and How to Fold a Hexagon

There is one last item to consider before gridding: the axes. I've explored parallel creases already. In pleat pattern terms, a grid is formed from two or more intersecting sets of parallel creases. An *axis* (plural *axes*) is one of those parallel crease sets. The most common grids are square and triangular grids. While the square grid is interesting, this text is dedicated entirely to the triangular grid and forms that can be produced on it. Many of these concepts are translatable to the square grid, so feel free to explore!

Figure 1.5.1 shows the square and triangle grid axes, the square grid having two axes (a horizontal and a vertical) and the triangle having three (oriented at 60° to one another).

Creating a grid amounts to one big process of *precreasing*, the act of creating folds and then unfolding them to prepare the paper with creases to be used later. The act of folding the paper along multiple previously-created creases is called *the collapse*.

In other words, the grid conditions the paper with a structured set of possible folds, making the pattern easier to fold later. Preparing a grid is especially useful for beginners due to the structure it provides, though experienced folders rarely omit the process due to the accuracy advantage, not to mention the visual interest it generates in the finished pattern.

Now that you know how to fold parallel creases, you can focus on arranging sets of these creases to create a grid. In order to create a structured form that's practical for consistent pleats, you must be deliberate in your choice of angle and maintain consistent spacing between the axes.

In Figure 1.5.2, each set of grids makes twists difficult to create for different reasons. In row A, the axes do not have the same spacing. In row B, the three axes of the triangle grid do not intersect at the same point. In row C, one of the axes in each example is not at an optimal angle to the others. In a square grid, you want 90° intersections and for a triangle grid, you want 60°–90° and 60° being whole number divisors of 360°.

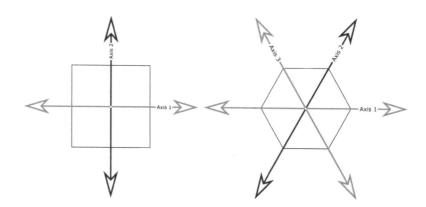

Figure 1.5.1 **Axes of a grid.**

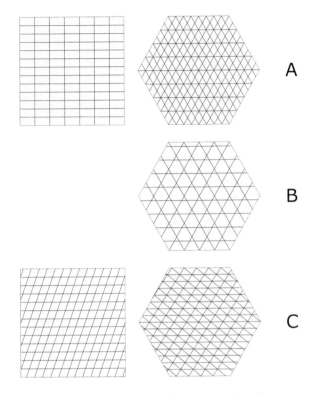

A

B

C

Figure 1.5.2 Grids that are difficult to work with.

How do you determine the axes of a grid so that the angle is consistent? Well, a square grid is relatively easy: every grid line is parallel or perpendicular to every other grid line. But if you don't have the guides provided by the paper edges, you will have to create the guides, either with a ruler, or with the perpendicular crease method shown in the last section.

To make sure the interval between pleats is equal throughout, you can either ensure that the outermost guides are the same distance apart (such as by starting with a square instead of a rectangle), or by measuring. You can also fold a diagonal to the grid, and if everything lines up, then the intervals are the same throughout, as shown in Figure 1.5.3.

But in a triangular grid, every crease is either parallel to or at 60° to every other crease on the paper. How do you construct one crease

at a 60° angle from another? The easiest way I've found is to start by constructing a hexagon, since the diagonals of a hexagon meet at its center at 60°.

To fold a hexagon, start with a rectangular sheet of paper. The images shown in the diagrams starting with Figure 1.5.4 use A4-size paper, roughly 21 × 30 cm. However, most rectangle proportions will work well for this construction, including the U.S. letter size of 8.5" × 11" (Figure 1.5.5).

Fold the paper in half lengthwise (Figure 1.5.5).

Now, rotate the paper so its raw edges are closest to you and fold one raw edge of the paper toward the folded edge farthest from you. The purpose of this is to find the ¼ mark of the paper. Then unfold that step (Figure 1.5.6).

Next, rotate again so the raw edge of the paper is farthest from you. Peel the right corner of the paper toward you so that the corner hits the ¼ mark made in Figure 1.5.6 and the crease formed also hits the folded corner closest to you. The resultant angle will be 60°. This is called a *pivot fold*, where you rotate a flap of the paper around a pivot point; in this case, that pivot point is the corner (Figure 1.5.7).

Now you need to find the 60° angle on the other side of the paper, so flip the paper horizontally and repeat the process for finding the ¼ mark (Figure 1.5.9).

Repeat the pivot fold with this side of the paper (Figure 1.5.10).

Once you have both pivot folds in place, you need to replicate those 60° angles. Bisect the angle farthest away from you (Figure 1.5.11).

Flip the paper. The lower left corner will be the center of the hexagon. Bring the right flap of the paper upward to bisect that angle (Figure 1.5.12).

At this point, you have all of the diagonals of the hexagon. The only thing left is to fold the top flap downward so that it passes through the rightmost corner, bisecting that corner's angle (Figure 1.5.13).

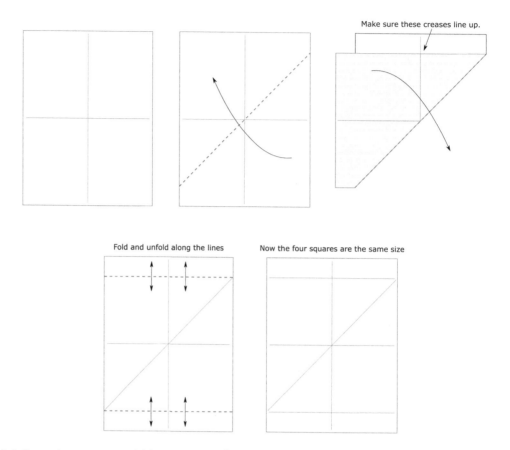

Make sure these creases line up.

Fold and unfold along the lines

Now the four squares are the same size

Figure 1.5.3 Preparing a square grid from a rectangle.

Figure 1.5.4 Starting rectangle.

Unfold and cut off the outside portions of the paper. Now you have a hexagon!

There are several methods for folding a hexagon from a square or rectangle, and you may want to investigate alternative approaches. The important goal is to attain three parallel, evenly-spaced creases along each axis and that the three axes approach at 60° angles to one another. The method previously described achieves this goal, with the addition of two ¼ mark creases used for its construction; you will use these creases in the gridding itself (Figure 1.5.14).

Using this method of hexagon folding, you end up with three creases with a mountain bias and two with a valley bias, or vice versa,

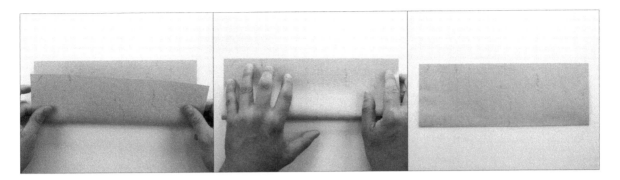

Figure 1.5.5 Fold in half lengthwise.

depending on which side of the paper you're viewing. To keep everything in line, it's good practice to backcrease every crease you've made by folding it first as a valley on one side of the paper and then as a valley on the other side of the paper. This ensures that every crease's memory is sufficiently neutral.

The backcreases have a memory of the last parity they were in. I call the side that was most recently folded with valley folds the "valley side" and the opposite side the "mountain side." The process of gridding will involve identifying valleys that stray onto the mountain side, and vice versa, and resolving their parity contradictions.

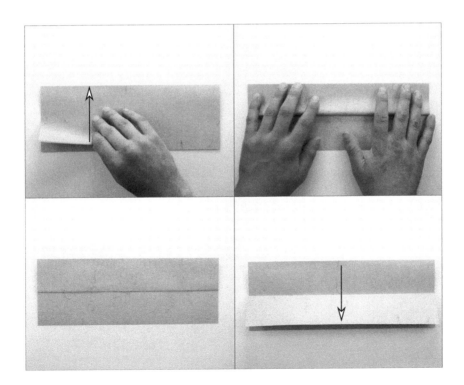

Figure 1.5.6 Fold one flap in half lengthwise and unfold.

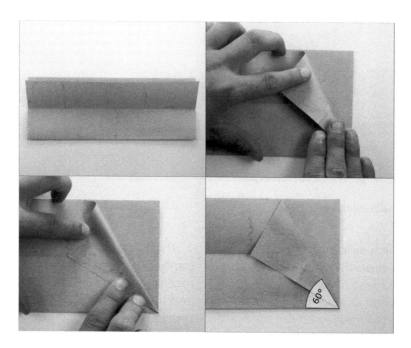

Figure 1.5.7 Pivot the corner of the flap to the crease made in Figure 1.5.6, with a crease that passes through the bottom corner. This will make a 60° angle.

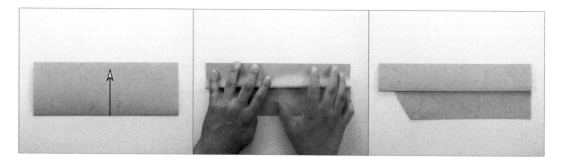

Figure 1.5.8 Flip and fold this side's flap in half lengthwise.

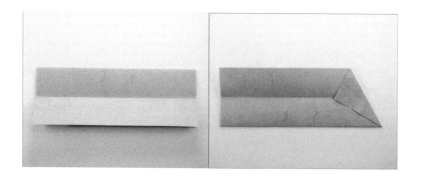

Figure 1.5.9 Repeat the pivot fold on this side.

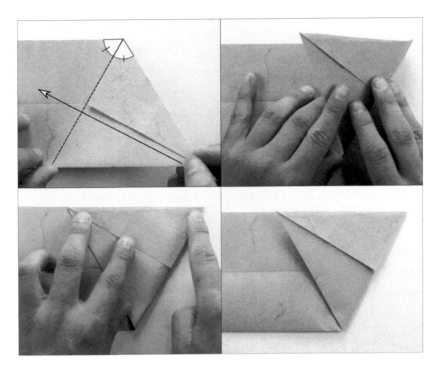

Figure 1.5.10 Bisect the angle in the top-right. Each angle will be 60°.

Figure 1.5.11 Flip the paper and bisect the lower-left corner.

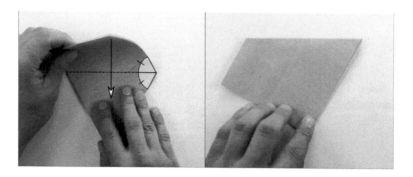

Figure 1.5.12 Bisect the corner on the side, folding the flap down through all layers.

Figure 1.5.13 Unfold and cut out the hexagon.

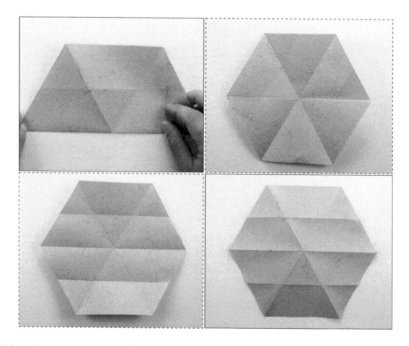

Figure 1.5.14 Refold each crease with one bias, and then backcrease each crease.

1.6 How to Fold a Triangle Grid

After preparing the hexagon, the paper will have three creases along the diagonals, running parallel to the hexagon's raw edges and intersecting at the center point. Therefore, if you accurately divide the intervals on each axis into halves, quarters, eighths, etc., you will end up with a grid that gets progressively more detailed and observes multiple identical intersections at identical spacings. This forms what is called a *triangle grid*, though it is also called a *hex grid*, due to its border shape. With an accurate triangle grid, your pleats will be predictable and easier to keep track of during folding.

Grids are often described according to their density, based on the number of divisions between opposite outer edges along one axis. With a hexagon made using the method previously outlined, two of the axes are folded into halves, while one of the axes is folded into fourths. The goal is to make each of the three axes identical in the number of divisions. The sequence I use for detailing a grid (from fourths into eighths, for example) is as follows:

1. Detail the first axis.
2. Backcrease the first axis.
3. Rotate and detail the second axis.
4. Backcrease the second axis.
5. Rotate and detail the third axis.
6. Backcrease the third axis.

You can repeat this process as far as you'd like, from fourths to eighths, 16ths, 32nds, 64ths, etc. This method doubles the number of triangles arranged along the diagonal of the hexagon for each level of subdivision, and each level of subdivision takes roughly twice as long as the previous level. For the diagrams in Chapter 1, you will need to subdivide to

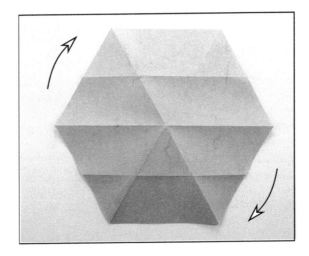

Figure 1.6.1 Rotate the hexagon.

16ths, meaning the diagonal of the hexagon will be comprised of sixteen triangle edges. For Chapter 2 diagrams, I recommend expanding to 32nds. The supplemental diagram at the start of Chapter 2 expands the techniques for gridding to 32nds.

To start the gridding, you've folded the hexagon, cut it out, and backcreased every fold at least once. Thanks to the hexagon folding method described previously, one of the axes is already subdivided into fourths, so we can leave it and subdivide another axis (Figure 1.6.1).

Rotate to an axis that is not yet subdivided into fourths, fold from the bottom edge of the paper to the parallel diagonal, and then unfold. Rotate 180° and repeat the same fold on the opposite side. You've now subdivided a second axis to fourths (Figure 1.6.2).

Flip the paper and backcrease the creases made in the last steps (Figure 1.6.3).

Flip back to the mountain side of the paper, and then rotate and repeat the detailing and backcreasing with the third axis. Now you have a "fourths" grid, and the foundation to continue. It is important to note when you finish the third axis of any subdivision that each of the crease intersections are the intersections of three creases,

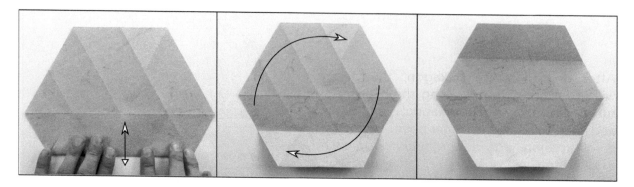

Figure 1.6.2 Second axis 4ths.

Figure 1.6.3 Backcreasing second axis 4ths.

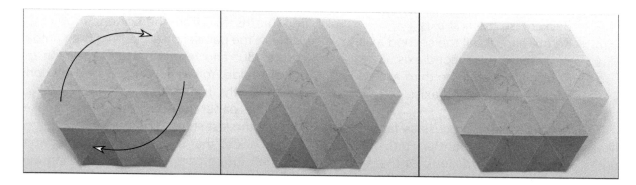

Figure 1.6.4 Third axis 4ths and backcreasing.

Figure 1.6.5 First axis 8ths start.

and one does not miss. It becomes more difficult as you subdivide further but is most important at the early stages of gridding (Figure 1.6.4).

To subdivide the paper into eighths along one axis, pinch the mountain fold at the one-quarter subdivision closest to the bottom edge of the paper, then bring that pinched crease to the parallel crease at the hexagon's diagonal. Press down and fold the three-eighths subdivision through the overhanging paper. Run your thumbnail along the hidden fold to prepare the next subdivision (Figure 1.6.5).

This creates the foundation for an accordioning stack of paper. Fold the overhanging edge on top of the stack to the parallel diagonal to subdivide into eighths. Then unfold, rotate 180° and create the ⅜ and ⅛ marks on the other side of the paper—or alternatively interpreted, the ⅞ or ⅝ marks from the first side of the paper (Figure 1.6.6).

Flip the paper to the valley side and backcrease the eighths folds made in the last steps. Remember, the goal is to make it so that every crease on this side of the paper was last folded as a valley. Then flip back to the mountain side (Figure 1.6.7).

Rotate and repeat the process of detailing the second axis to eighths using the same stacking method. Flip and backcrease, and then flip it back to the mountain side. Then, repeat with the third axis to finish the detailing to eighths (Figure 1.6.8).

After folding eighths, the process repeats iteratively, each level of detailing adding more evenly-spaced creases, and more layers to the accordion stack you create (Figures 1.6.9–1.6.11).

As you are backcreasing, pay special attention to the crease closest to the edge of the paper. If the paper is going to tear, this is the most likely place for it to do so (Figure 1.6.12).

After you've folded and backcreased all three axes, you've finished your 16ths triangle grid! Can you feel the difference between the mountain and valley sides of the paper?

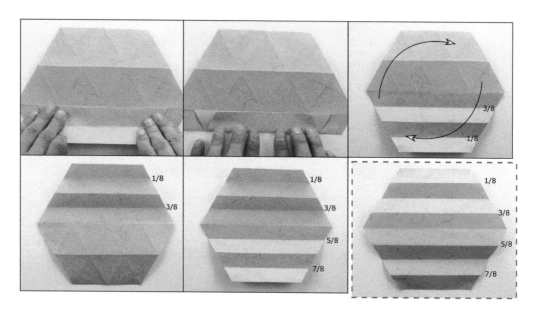

Figure 1.6.6 **First axis 8ths.**

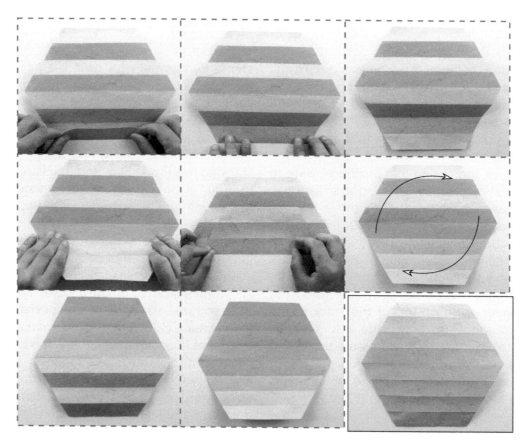

Figure 1.6.7 **First axis 8ths backcreasing.**

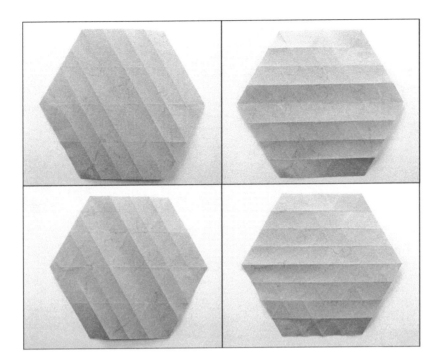

Figure 1.6.8 Second and third axes 8ths.

Figure 1.6.9 First axis 16ths and starting a stack.

Figure 1.6.10 First axis 16ths and continuing the stack.

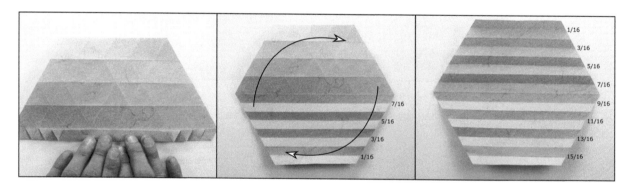

Figure 1.6.11 First axis 16th finished.

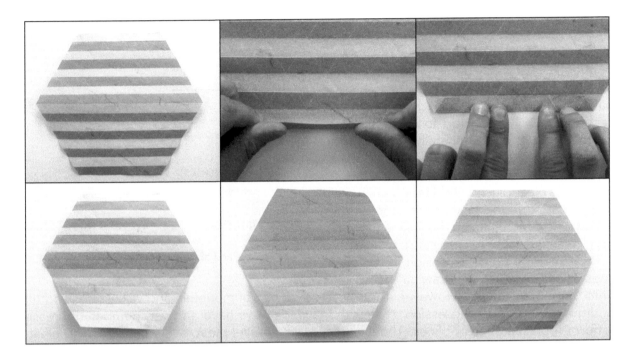

Figure 1.6.12 First axis 16ths backcreasing.

Figure 1.6.13 Second and third axes 16ths finished.

1.7 Simple Pleat

At the core of pleat patterns is the pleat. Most frequently used in this book will be the *parallel pleat*, composed of two creases of opposite parity that are parallel to each other. In this section, we will use a grid to make a single-width parallel pleat. I generally start with the mountain side of the grid facing toward me as it is easier to pinch that side.

To fold a pleat, pinch a grid line into a mountain fold and lift it off the table. Lay that fold down flat away from you, turning the adjacent parallel grid line into a valley fold (Figure 1.7.1).

Pleats can be used in a design to represent lines, just like you can draw or paint lines onto paper or canvas. However, they have a different type of expression than other artform representations of lines. They can be wide, thin, complex, curved, hidden, or three-dimensional. They can lie down on one side or the other, which creates a visual difference that seems to "point" in a particular direction. The way pleats intersect with other pleats is more complicated than the way drawn lines intersect with other drawn lines, due to their material and mathematical constraints. The pleat is the building block for all the complexity put forth in this book.

A pleat is composed of mountain and valley folds. A flat-folded pleat is composed of at least

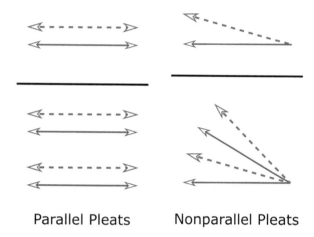

Parallel Pleats Nonparallel Pleats

Figure 1.7.2 Parallel versus nonparallel pleats.

one pair of mountain and valley folds and thus will always have an even number of creases. In a *parallel pleat*, every crease in the pleat is parallel to the others. Conversely, the creases of a *nonparallel pleat* will eventually intersect if nothing else interacts with them first.

A *mono-pleat* has only a single mountain and valley fold, whereas a *composite pleat* has two or more pairs of mountain and valley folds.

The pleat type I will be using the most in this text is the *simple pleat*, a pleat composed of only a single mountain and valley fold, with both of the creases being parallel to each other. In other words, a simple pleat is one that is mono and parallel.

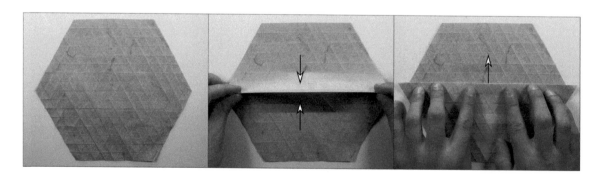

Figure 1.7.1 Creating the simple pleat.

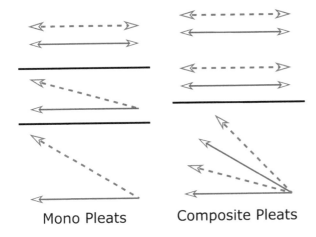

Mono Pleats   Composite Pleats

Figure 1.7.3 Mono versus composite pleats.

Simple Pleat

Figure 1.7.4 A pleat that is mono and parallel is a simple pleat.

Reorienting a Pleat

A pleat that is flat is said to be *oriented* in one direction or the other. The *orientation* of the pleat is determined by which side of the mountain fold the valley fold is on.

To change the orientation of the pleat, lift the pleat off the table and flip it in the opposite direction, like a switch. Lay the pleat flat, and its orientation is reversed (Figure 1.7.5).

Although the simple pleat is composed of a mountain and a valley fold, the reflection of the valley fold in the overhanging paper has an important role to play, as reorienting shows. This reflection can be called the *valley shadow*, or just the *shadow*. The shadow is not an actual fold, but rather represents the alternate location for the valley fold should the pleat be reoriented. Reorienting switches a pleat's valley fold to its shadow, and vice versa.

Once you've folded a pleat and reoriented it, the paper will have obtained a strong enough memory that after unfolding (and even flattening out), the mountain fold will likely stand out among the rest of the grid.

I call this the *standing* form of a pleat. This allows you to clearly see the mountain memory,

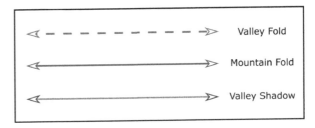

Valley Fold

Mountain Fold

Valley Shadow

Figure 1.7.6 Valley fold, mountain fold, and valley shadow of a simple pleat.

Figure 1.7.5 Reorienting a simple pleat.

Figure 1.7.7 Simple pleat in standing form.

as well as each of the two options for its valley folds. This accentuates where you've already folded and helps you avoid getting lost in later sections when there are more interacting pleats.

Drifting a Pleat

Reorientation changes a pleat into an alternate form; it doesn't change the location of the pleat. Another useful folding skill to acquire is called *drifting*; this is the act of translating a pleat parallel to itself several units in a direction. Like reorientation, drifting can affect a pleat in either direction, parallel to the pleat's original location. To drift a pleat one unit, the folder displaces the mountain and valley fold of the original pleat one parallel grid space away and then flattens the displaced pleat along the length of the paper.

To do this, begin with the pleat you want to drift in standing form.

Choose which direction you want to drift the pleat; in the illustrated example in Figure 1.7.8, I am drifting one grid space below the standing pleat—closer to myself. Place one finger underneath the valley fold that you intend to backcrease and turn it into a mountain fold. Work your way along the length of the pleat, manipulating it so the valley fold becomes a mountain fold all the way across. The standing pleat's mountain fold will turn into a valley fold as a result, and its other valley will become unfolded. At the end of this maneuver, the original pleat will have shifted one grid space away to a new location on the grid.

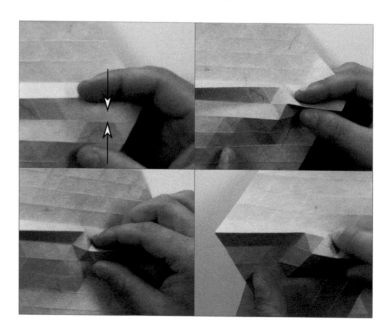

Figure 1.7.8 Drifting process.

Figure 1.7.9 Finished drift. LEFT: standing form. MIDDLE and RIGHT: different orientations.

To show the new location of the pleat, this is its new standing form in Figure 1.7.9. Again, you should clearly be able to see where the pleat ended up, one space away from its original location.

Splitting a Pleat

Splitting a pleat is a more complex maneuver and is useful for learning how to manipulate multiple pleats.

Begin with a simple pleat on a triangle grid in standing form and choose a point on the mountain fold. In this case, you'll choose the hexagon's center point. From this point, the pleat extends in two directions. When you split the pleat, one side around the chosen point stays intact. The other side is unfolded to introduce new pleats. To execute this maneuver, pinch the pleat on the side that will stay intact, and then, placing your other hand on the backside of the paper, undo the other side of the pleat until you reach the chosen split point (Figure 1.7.10).

From the chosen point, pinch mountain folds that extend from that point toward the edges of the paper. In this case, the new mountain folds extend from the center point of the hexagon to its corners, so that every other corner of the hexagon will have a mountain pinch running toward it (Figure 1.7.11).

Lay those new mountain folds flat, oriented away from the first pleat. This is called a *pleat split*. I will call the initial pleat the *parent pleat* and the new pleats the *children pleats*.

You can reorient all three pleats once they intersect. Play with the reorientation of one or more pleats and see what the effect on their intersection is.

Figure 1.7.10 Undoing the pleat.

Figure 1.7.11 Forming the mountain folds of the children pleats.

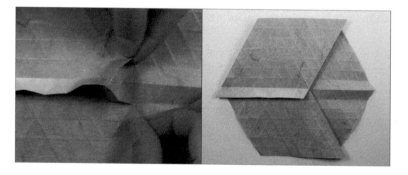

Figure 1.7.12 Forming the valley folds of the children pleats.

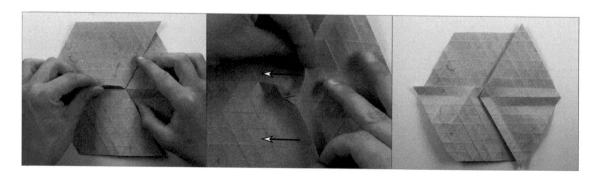

Figure 1.7.13 Reorienting the children pleats.

1.8 The Six Simple Twists

This section delves into this text's namesake: the Six Simple Twists. I have chosen these six for their ubiquity in origami tessellations, as well as to encourage the reader to apply the maneuvers and techniques described previously in Chapter 1.

However, before I get into that I should discuss what a twist is. In the most general sense, a twist is the result of a *pleat intersection*. A pleat intersection is the intersection of multiple pleats, though they do not always meet at a point. I often study how the pleats intersect, whether they do meet at a point or if one of them is offset from the others. The piece folded in Figure 1.7.13 is an example of a pleat intersection. Though it was created from the center outward, it is helpful to think of the pleats as coming from the edge of the paper toward the intersection. I document a pleat intersection with a modified CP, which I will call a *pleat schematic*.

To create the pleat schematic, draw the mountain folds from the edge of the paper each at the same rate. Stop a mountain fold when it hits another mountain fold. Then draw the valley folds parallel to the mountain folds from the edge of the paper. Stop a valley fold when it hits a mountain fold—it can pass through another valley fold. Then, I generally like to draw a circle encompassing every vertex in the design. This allows the viewer to focus

more on the pleats than the intersection, and in a pleat schematic, the pleats are the important part.

The pleat schematic describes the approach of the pleats, which gives you clues as to the structure of the eventual twist, but does not give you information of what happens at their intersection (inside the circle). I call the area where the pleats converge a *molecule*, a piece of a full origami tessellation. There are multiple possible molecules that can be created from a given pleat intersection, but each molecule is created by one pleat intersection. *Simple molecules* are molecules composed of pleat intersections with simple pleats.

Twists are molecules where, in order to lie flat, a layer of the paper rotates as the molecule is formed. Not all molecules rotate as they are folded flat, which is why twists are a more specific form of a molecule. *Simple flat twists*, or just simple twists, are twists that are created from pleat intersections with simple pleats and lie flat [1]. Simple twists are thus a subset of simple molecules and can often be folded by starting with simple pleats converging in a specified way.

With apologies, I acknowledge that the title of this book is slightly misleading. The twists described here are simple from a mathematical perspective. That does not mean that they are easy to fold or understand for beginners, though they do get easier with practice. There is no one proper way of folding a twist;

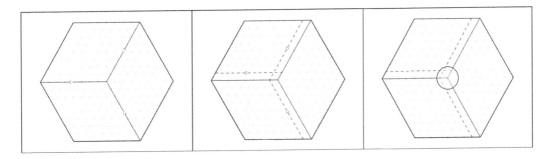

Figure 1.8.1 Forming a pleat schematic.

different folders have different methods that they find more intuitive. For this reason, I have included multiple folding methods for several of the twists described. For each twist, try to identify which sections of the paper are being brought together, the width of the pleats, and the nature of the twist that occurs to gain an understanding of its mechanics.

1.9 Triangle Twist

Triangle Twist Method One

The triangle twist is a natural extension of pleat splitting. As you played with reorienting the pleats, you may have noticed that when each of the pleats is oriented in the same

rotational direction (all CW or CCW), the paper resists slightly and does not fold flat. Like the standing form of a pleat, we will call this the *standing form* of the twist. This is the state where all pleats are in place, but the twist itself has not yet been flattened.

By pushing through this resistance, the paper in the middle spreads apart and, when fully flattened, creates a triangle platform. This is called the *triangle twist*. This twist introduces new creases not originally on the grid. The grid lines are merely there to guide the triangle twist's formation.

To assist in developing your understanding of this twist, I have added the CP to Figure 1.9.3.

Triangle Twist Method Two

Splitting a pleat, orienting the pleats in the same direction, and flattening the standing form offers one method for creating a triangle twist, but it's good to know alternate methods.

For this second method, start by pinching every other diagonal of the hexagon as a mountain fold, and walk the mountain folds in standing form toward the center, pinching along the way.

When you reach the center, the pleats will snap into place and can be flattened all in the same direction, creating the standing form of the triangle twist you saw in the last method. You can flatten the twist from there the same way.

Figure 1.9.1 Standing form of the triangle twist.

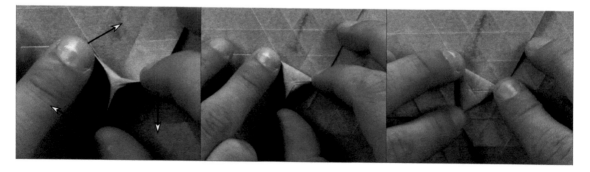

Figure 1.9.2 Flattening the twist.

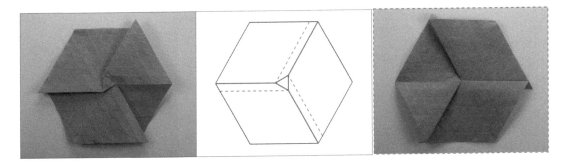

Figure 1.9.3 LEFT: Finished triangle twist front. MIDDLE: CP of Triangle Twist. RIGHT: Finished triangle twist reverse.

Figure 1.9.4 Pinching the mountain folds of the pleats.

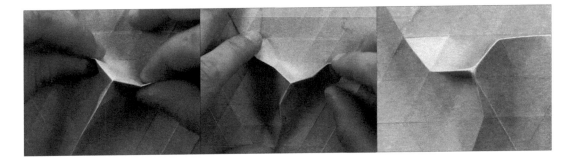

Figure 1.9.5 Walking the pleats toward the center.

1.10 Triangle Spread Twist

The triangle spread twist is an excellent example of how you can use drifting to change the form of a pleat intersection. You can drift one pleat away from the pleat intersection to create a triangle spread twist.

Start with a standing triangle twist in the center of the hexagon with a CCW rotation.

Drift one of the pleats, displacing it one unit CCW of its current position. Keep pinching the new pleat toward the center until it pops into a triangular platform in the (relative) center of the paper. This is the standing form of the triangle spread.

Press down on the corners of that triangle, and the platform will spread and flatten,

Figure 1.10.1 Standing triangle twist.

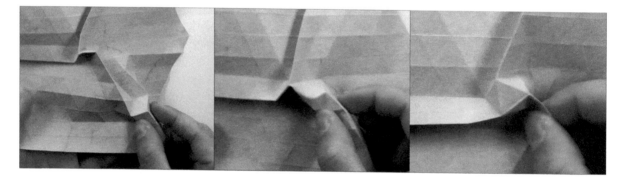

Figure 1.10.2 Drifting one pleat CCW.

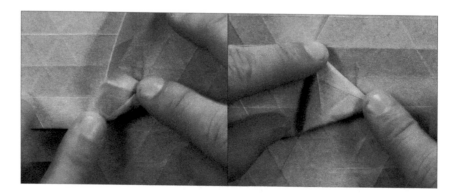

Figure 1.10.3 Flattening the twist.

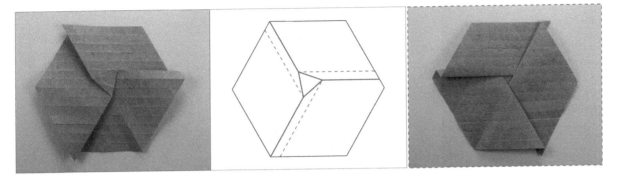

Figure 1.10.4 LEFT: Triangle spread twist. MIDDLE: CP of the triangle spread. RIGHT: Triangle spread reverse.

similar to the triangle twist before (this will be a common action throughout the book).

This is called the *triangle spread twist* or *open back triangle twist,* the latter name being due to the triangular "negative" space on the reverse of the twist.

1.11 Hex Twist

Hex Twist Method One

I begin most of my patterns with the hex twist, and it holds a special place in the hearts of practitioners of this craft due to its flexible use as a starting point for origami tessellations. New folders tend to find this twist unintuitive, however, because it can be challenging to set the pleats in place, and because a surprising amount of rotation is required for the twist to lie flat. I offer two methods here that will hopefully allow you to push through this learning curve. Take your time and allow yourself to fail a few times before you understand this important twist.

I like to start by folding and unfolding the pleats that are going to go into the twist, one at a time. For the hex twist, create a standing pleat on one of the diagonals so you see the memory of the pleat clearly on the paper. Rotate 60° and repeat, until the remaining two axes are pleated into the paper's memory.

Even though you folded three pleats that run straight through the diagonals, it is helpful to think of these as *six* distinct pleats, each one approaching the center from the corner of the hexagon.

Figure 1.11.1 Pinching the mountain folds of the pleats.

Figure 1.11.2 Setting up the collapse.

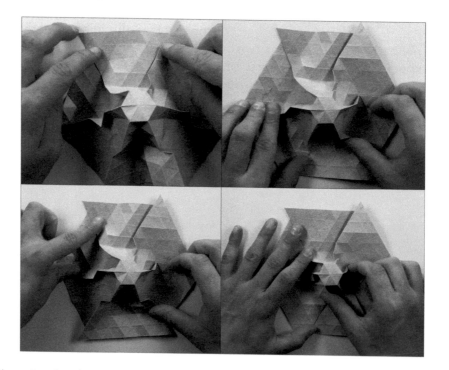

Figure 1.11.3 Flattening the pleats.

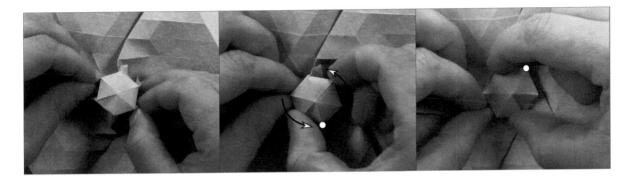

Figure 1.11.4 The twist.

Pinch two opposite diagonal pleats at the edges and walk your fingers toward the center, pinching all the way. When you reach the center, notice the hexagonal platform that rises toward you where the six pleats meet.

Start to lay some of the six pleats flat, as well as you can, in a CCW fashion from the corners of the paper inward. Work the pleats flat until the center hexagonal platform is the only paper not lying flat on the table.

Rotate the central hexagon CCW until the entire platform lies flat. The rotation will be greater than you would expect the first time you fold this twist. Follow the white dot in the second and third photos of Figure 1.11.4 to see the rotation; the hexagonal platform rotates 120°.

Figure 1.11.5 shows the finished *hex twist*.

Hex Twist Method Two

Given that the hex twist is rather complicated, if you are having difficulty, you might try this alternate method of folding it. Like before, fold a standing pleat on each of the diagonals of the hexagon. Pinch two opposite pleats until the center hexagon reveals itself. Push on it to accentuate.

Bring two opposite corners of the hexagon toward each other and have them pass until they can't be pushed any farther.

This creates a flat form. Peel apart the rest of the paper from the reverse side and lay the paper flat, refolding one of the valleys of the standing pleat that is currently the top folded edge.

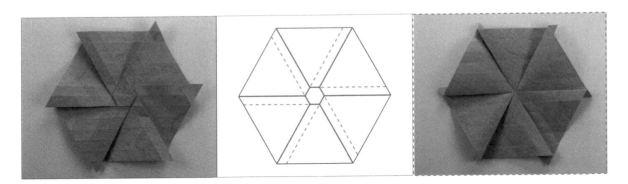

Figure 1.11.5 LEFT: Hex twist front. MIDDLE: Hex twist CP. RIGHT: Hex twist reverse.

Figure 1.11.6 Pinching the mountain folds of the pleats.

Figure 1.11.7 Connecting the corners of the hexagon.

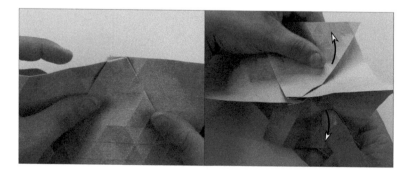

Figure 1.11.8 Peeling the hexagon apart.

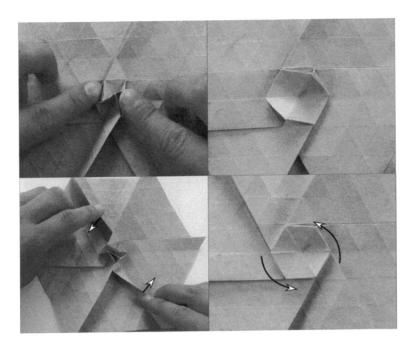

Figure 1.11.9 Opening the twist.

This results in the hexagonal platform in the middle, which was your goal. However, the pleats are oriented in different directions, and the twist might be rotated in the wrong fashion. Reorient the pleats until they all lay flat CCW. Then rotate the hexagon CCW as far as it can go.

1.12 Hex Spread Twist

The relationship between the triangle spread and the triangle twist is analogous to the relationship between the hex spread and the hex twist. To create the hex spread, partially fold CCW-oriented pleats with mountain

Figure 1.12.1 Pinching the mountain folds of the pleats.

41

Figure 1.12.2 Standing form of the hex spread.

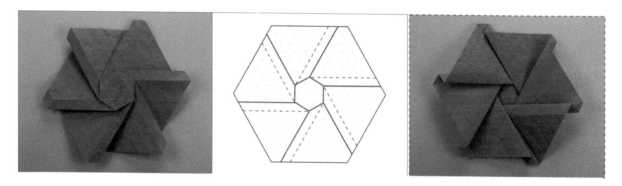

Figure 1.12.3 LEFT: Finished hex spread twist front. MIDDLE: CP of hex spread twist. RIGHT: Hex spread twist reverse.

folds, displaced one grid line CCW from the hexagon diagonals. One at a time, fold each pleat to the center and then unfold. When you've finished, the paper will retain the memory of all six pleats in the paper. Pinch all three pairs of opposite pleats and lay them flat to reveal a hexagonal star in the center of the paper.

The star is the standing form of the hex spread twist. Push the interior corners of the star toward the table to finish the twist.

This is called the *hex spread twist*, or *open back hex twist*, the latter name being due to the hexagonal "negative" space on the reverse of the twist in the same way the triangle spread had a triangular space on the reverse.

1.13 Rhombic Twist

The rhombic twist is best thought of as a mechanism for creating a sudden translation of a pleat, as you will see in Chapter 2. As a pleat travels across the paper, it can be useful to shift the pleat in one direction or the other. The rhombic twist allows that.

As with the other twists, you'll start with folding standing pleats along the diagonals of the hexagon. Here, only do this on two axes. Consider these to be the intersection of four distinct pleats—rather than two—running from the edge to the center of the paper.

Notice that there are two sets of consecutive axes with pleats on them. Each of these sets

Figure 1.13.1 Pinching the mountain folds of the pleats.

has one pleat that is most CCW and one that is most CW. I will call the pleat that is most CCW of each set the *leading pleat* and the pleat that is most CW of each set the *trailing pleat*.

To form the rhombic twist, you have to drift one of the leading pleats CCW (if in a CCW rotation) or one of the trailing pleats CW (in a CW rotation) one unit.

As you form the drifted pleat toward the center of the paper, you'll notice a rhombus forming where the pleats converge. Flatten the twist by pushing on the rhombus's obtuse corners.

This forms the finished *rhombic twist*.

When folding the rhombic twist, it is easy to forget which pleat to drift. In Figure 1.13.4 the first figure shows the correct drifting for a CCW rhombic twist, and the second figure shows a drift that will be more troublesome. The subsequent photos show what the physical result is of the wrong drift. If your twist looks like this, undo the drift and drift the other pleat instead.

Rhombic Twist Method Two

An alternate method for setting up the pleats is a little simpler. Start by folding a pleat across the diagonal; do not unfold. Then, fold a second pleat through the first one. Reorient that second pleat to put it in standing form, then unfold fully.

Pop the corners of the marked rhombus so the corners are all mountain points and the rhombus lifts up. No drifting is required, and you can lay the pleats flat to create the rhombic twist.

Figure 1.13.2 Standing rhombic twist.

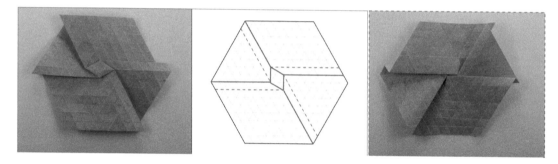

Figure 1.13.3 LEFT: Finished rhombic twist front. MIDDLE: CP of rhombic twist. RIGHT: Finished rhombic twist reverse.

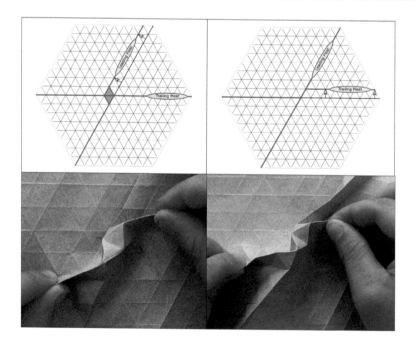

Figure 1.13.4 Attempting to drift the wrong pleat. TOP-LEFT: Schematic for drifting the leading pleat. TOP-RIGHT: Schematic for drifting the trailing pleat. BOTTOM ROW: Photos of drifting the trailing pleat.

Figure 1.13.5 Locking two pleats.

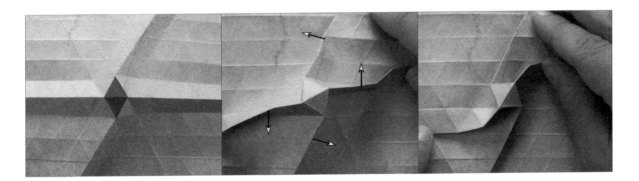

Figure 1.13.6 Standing form of the rhombic twist.

1.14 Arrow Twist

This twist is formed by the intersection of four pleats in an arrangement of 60°, 60°, 120°, and 120° angles. What makes this twist different is that one of the pleats is twice as wide as the others. I'll explore why more in Chapter 3, but for now keep that in the back of your mind while folding.

Begin with a split pleat (pleat oriented as you choose, but do not flatten it to a triangle twist). Fold a pleat through the split along the

45

diagonal. This will cause one of the original pleats of the split pleat to double in width. Reorient each of the four pleats, including the new double-width pleat, so that they all lie CCW around the center.

You'll notice that one of the pleats is trapped inside the other three. Gently pull the trapped pleat to release the paper and lay all pleats flat CCW again to reveal a platform that looks the same as the rhombic twist. This is the standing form of the arrow twist.

Flatten the rhombus platform to create the full twist. I call this an *arrow twist* due to the arrow formation of the three pleats separated

Figure 1.14.1 Grafting a pleat through a three-wasy intersection.

Figure 1.14.2 Standing form of the arrow twist.

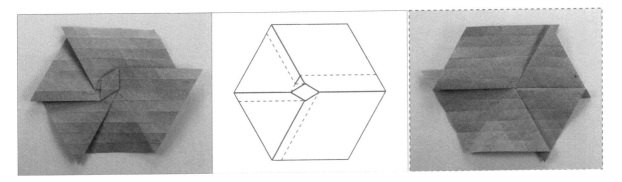

Figure 1.14.3 LEFT: Finished arrow twist front. MIDDLE: CP of arrow twist. RIGHT: Finished arrow twist reverse.

Figure 1.14.4 A notch.

by 60° angles. It's similar in form to the rhombic twist, in that the platform is the same and there are four pleats. The differences lie in the angles

of the pleats in relation to each other and the fact that one of the pleats is double-width.

Additionally, extra paper is displaced that looks almost like another twist beneath the first. I call this extra hidden twist a *notch*. I'll explore notches more in Chapter 3, so it's worth recognizing this effect early in the learning process.

Arrow Twist Method Two

Like the other twists, it's helpful to be able to parse the arrow twist as an approach of pleats toward the center. For this method, remember that there are four pleats, three from three consecutive corners, each single-width, plus a fourth double-width pleat that extends opposite the middle pleat of the single-width pleat cluster.

Figure 1.14.5 Pinching the mountain folds of the pleats.

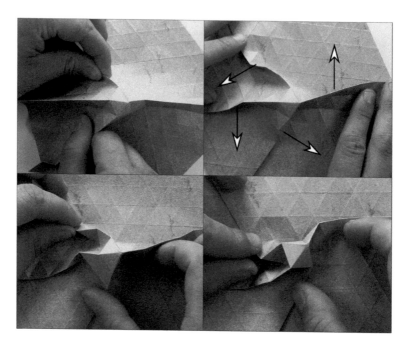

Figure 1.14.6 Creating the valley folds of the pleats.

Walk the pleats inward, flattening CCW, until they lay flat and the twist's standing form's rhombic platform becomes apparent. Flatten the twist as instructed previously.

1.15 Anatomy of a Molecule

With these six twists, you have some of the tools required to advance to full-scale patterns.

To better understand the direction this book will take in Chapter 2, Figure 1.15.1 shows pleat intersections and gives a glimpse of the tilings you'll begin to create. The left part of that figure represents a single pleat intersection; in this drawing, there is only a single molecule on the paper. Making smaller pleat intersections—such as from a smaller grid size—results in being able to fit several on the same sheet of paper as shown in the

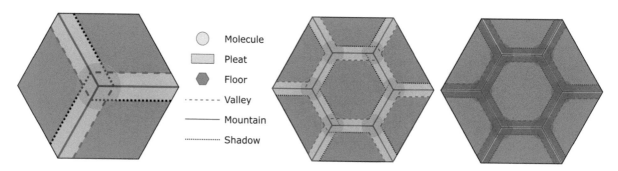

Figure 1.15.1 LEFT: Single-pleat intersection. MIDDLE: A common format for tiling intersections. RIGHT: Reverse or floor side.

middle drawing. The result is an origami tessellation. The reverse of this form shows mostly the spaces between the pleats, which I will call the *floors*.

1.16 Pleat Intersection Notation

To create a triangle twist, I started with a simple pleat and split it. To create a triangle spread twist, I set up the same pleats required for the triangle twist and drifted one of the pleats away from the intersection. The mountain folds of the spread pleats still intersect, but not at a single point anymore. The same holds true for the difference between the hex and the hex spread (with more than one pleat drifting in the latter case).

While both twists have similarities, it helps to be able to describe the precise differences between them. In this section, I will introduce a notation system initially proposed by Matthew Benet and jointly developed for this purpose; I will call it the *Benet Notation*—named after its initial proposer. The Benet Notation is adaptable to the square grid, but I will only

work with it in the context of the triangle grid on a hexagon. For now, all you need to understand is how this notation can describe pleat intersections.

I will further explore this system and its mathematical implications in Chapter 3; at this point, the notation simply offers a systematic way to document the pleat intersections you've just learned. When folding a grid, you should think of the hexagon as having three axes. However, since in a pleat intersection, pleats coming from opposite sides of the hexagon can be drastically different, it is helpful to think of the grid as having six axes when describing pleat intersections instead of just three.

Benet notation begins with a gridded hexagon as shown in Figure 1.16.2. Recall that the hexagon's diagonals are called the *axes* of the grid. The axes intersect at the hexagon's center. We will name these axes $A_1, A_2, A_3, A_4, A_5,$ and A_6, starting from the right-most diagonal, rotating CCW. Every grid line parallel to an axis is given a number, based on its distance in grid spaces from the axis. The number is a positive integer if the grid line lies on the CCW side of the axis and a negative integer if it lies on the CW side.

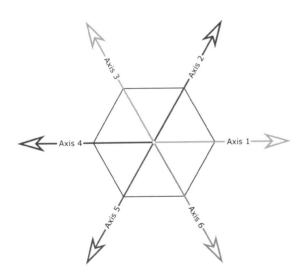

Figure 1.16.1 Six axes of a pleat intersection.

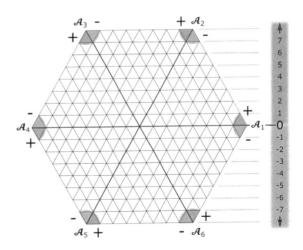

Figure 1.16.2 The Benet coordinate system.

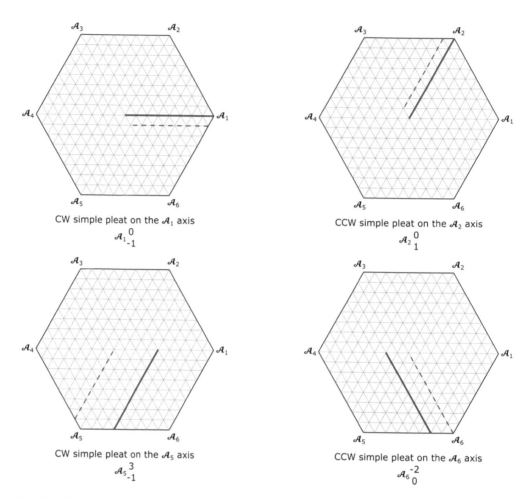

Figure 1.16.3 Simple pleat notation and pleat schematics.

Any simple pleat that extends toward the center can be described using the form A_{nv}^{m}, where A_n is the axis parallel to the pleat, m is the location of the mountain fold in the pleat, and v is the location of the valley fold in the pleat; both m- and v-values are the number of grid spaces away from the diagonal of A_n.

A positive m results in a pleat where the mountain fold is CCW of the axis, and a negative m results in a pleat where the mountain fold is CW of the axis. Likewise, a positive p results in a pleat where the valley fold is CCW of the axis, and a negative p results in a pleat where the valley fold is CW of the axis.

Pleats do not generally just stop when they reach the center of the paper or the neighboring axis, but they also don't usually travel through an intersection without changing in some way. To fully describe a pleat intersection, you can use an ordered hextuple, which shows the A_{nv}^{m} form for each of the six axes, starting at A_1 and ending at A_6, each pleat separated by commas. If there is no pleat approaching from that axis, you put a Ø, read as "null." In Figure 1.16.5 you will find the notation for each of the six simple twists, with their corresponding CPs.

Figure 1.16.4 Simple pleat examples.

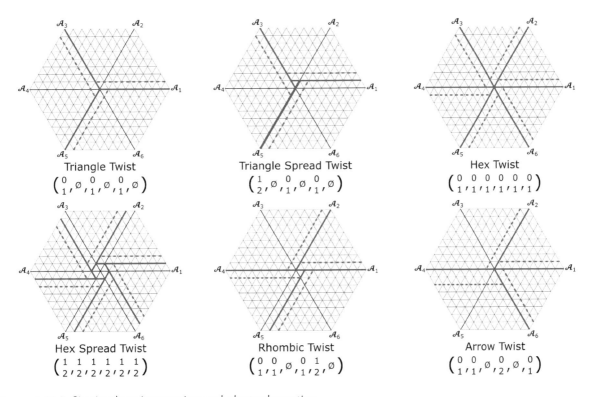

Triangle Twist
$$\begin{pmatrix} 0 & 0 & 0 & 0 \\ 1 & \emptyset & 1 & \emptyset & 1 & \emptyset \end{pmatrix}$$

Triangle Spread Twist
$$\begin{pmatrix} 1 & 0 & 0 & 0 \\ 2 & \emptyset & 1 & \emptyset & 1 & \emptyset \end{pmatrix}$$

Hex Twist
$$\begin{pmatrix} 0 & 0 & 0 & 0 & 0 & 0 & 0 \\ 1 & 1 & 1 & 1 & 1 & 1 & 1 \end{pmatrix}$$

Hex Spread Twist
$$\begin{pmatrix} 1 & 1 & 1 & 1 & 1 & 1 & 1 \\ 2 & 2 & 2 & 2 & 2 & 2 & 2 \end{pmatrix}$$

Rhombic Twist
$$\begin{pmatrix} 0 & 0 & 0 & 1 \\ 1 & 1 & \emptyset & 1 & 2 & \emptyset \end{pmatrix}$$

Arrow Twist
$$\begin{pmatrix} 0 & 0 & 0 & 0 \\ 1 & 1 & \emptyset & 2 & \emptyset & 1 \end{pmatrix}$$

Figure 1.16.5 Six simple twist notation and pleat schematics.

Chapter 2

How to Use the Six Simple Twists

The twists you learned near the end of Chapter 1 are pieces to a fuller puzzle. Their role in origami tessellations becomes clearer when you arrange them in evenly-spaced multiples on the paper. Chapter 2 takes these building blocks and presents techniques for combining them to form patterns. Before continuing, I recommend practicing the six simple twists, especially the triangle and hex twist.

2.1 32nds Grid

Going forward, you will want more triangles on the grid. Each pleat (twist, etc.) uses up a certain amount of paper, and a grid of 16ths can only fit so many before you run out of space. A denser grid means thinner pleats, thus smaller twists, and more free paper on the grid. This free paper could be used for more negative space, more twists, and—this is where things get exciting—more complex designs!

To detail a grid from 16ths to 32nds, I will use the method described in Chapter 1: detail the first axis, backcrease the first axis, rotate and detail the second axis, backcrease the second axis, rotate and detail the third axis, backcrease the third axis. More detailed grids present their own challenges, and this section offers techniques for getting through them.

The first time you fold a dense grid, it's unlikely that all of your creases will line up perfectly. That's alright! Accuracy comes with time and practice. In the meantime, work with whatever you end up with. When folding a complex grid, I recommend some flowing music, something with a good beat. Get into the rhythm of it, and don't focus too much on the goal but rather enjoy the process. You can fold the grid quickly or slowly, but when you are done, you will have prepared your canvas and are ready to explore this craft more fully.

Figure 2.1.1 Folding very small subdivisions.

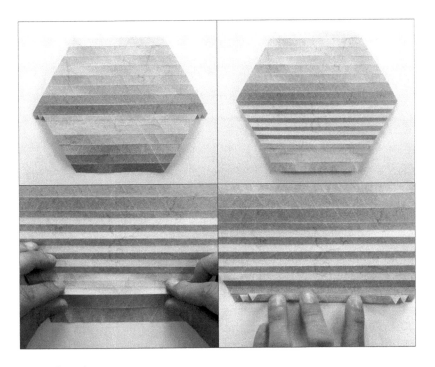

Figure 2.1.2 Halting a stack and restarting.

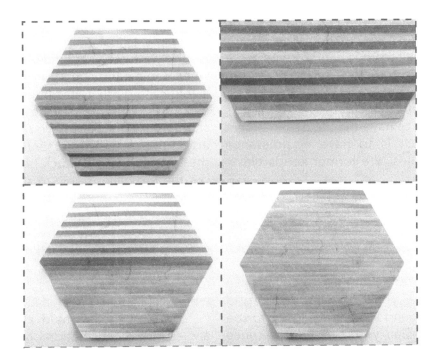

Figure 2.1.3 Backcreasing first axis 32nds.

Figure 2.1.4 LEFT: Finished 32nds valley side. RIGHT: Finished 32nds mountain side.

As you fold thinner and thinner subdivisions, you are more likely to accidentally curve a crease and form a grid line that isn't parallel to the others. Be careful to ensure the creases remain straight. Keep your fingers tight on the mountain pinch until they reach the edge of the paper, and then work your way on the other side of the crease. If you do start to curve, you do not have to redo the entire grid. Just straighten the curved crease and go on to the next one. When creating your stacks, if you feel yourself losing accuracy with more layers, just unfold the stack and restart from where you left off (Figure 2.1.2). Should you find yourself with inaccurate gridding, it is more important to have the creases intersect at points than it is to have equidistance between grid lines; the former keeps twists more in line.

As the grid gets more detailed, pay even more attention when backcreasing closer to the paper's edge. The crease closest to the edge becomes more likely to tear with each level of detailing.

If you get lost, identify which axes you have done. Sometimes you'll have to lift the grid off the table or bend the paper slightly to see where you are. Can you feel the difference between the mountain and valley sides of the paper?

2.2 Locking and Unlocking Pleats

There is one last skill to learn before you fold your first full pattern. I call this *pleat locking*, and its counterpart action, *pleat unlocking*. Two pleats are *locked* when they cross each other. The pleat folded last *locks* the one folded first in place so that it cannot be unfolded without first unfolding the top pleat.

Two pleats that intersect in this way will often result in a twist in the final design. However, when one pleat is already in place and another is folded through it, it is often easier to simply lock them temporarily so that the pleats stay under control until you are ready to pay attention to their intersection. Notice how the bottom pleat is stuck beneath the other, but the top pleat is able to stand and reorient without issue (Figure 2.2.1).

Pleats coming from molecules will often become locked together when they intersect; consequently, it is useful to know how to unlock two pleats when the ends of those pleats are tethered down. To represent this, I have attached binder clips to the paper in Figure 2.2.2.

To unlock pleats, hold along the obtuse angles of the intersection on each side of the intersection and pull the intersection apart, moving your hands directly away from one

Figure 2.2.1 Locking two pleats.

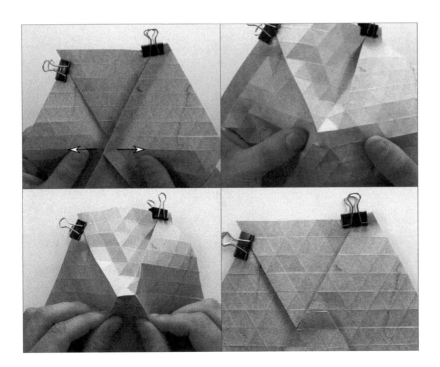

Figure 2.2.2 Unlocking two pleats.

Figure 2.2.3 Unlocking into a bar. TOP ROW: Bar of width one. SECOND ROW: Bar of width two. THIRD ROW: Bar of width three. FOURTH ROW: Bar of width four.

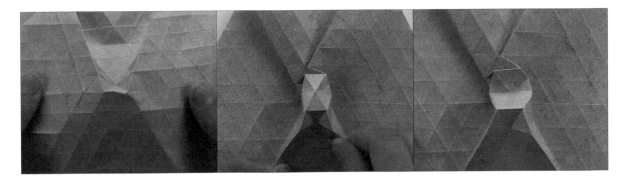

Figure 2.2.4 Setup for unlocking into a hex twist.

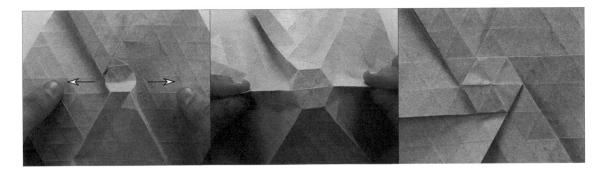

Figure 2.2.5 Unlocking into a hex twist.

another to release the trapped paper. Reform the pleats you pulled apart, and the displaced paper resolves into part of a hex twist. At this point, the pleats are unlocked and can be further worked on.

Bar Unlocking

There are different ways to unlock pleats. I'll call the unlock shown in Figure 2.2.2 a *bar unlock* with a width of one. By opening the pleats further, I can change the width of the bar unlock, as seen in Figure 2.2.3.

Bar unlocking gives you flexibility to place the pleats after the unlock.

Unlocking into a Hex Twist

Sometimes it helps to unlock pleats directly into a hex twist.

To do this, unlock into a bar with a width of one. Pinch the mountain folds that extend the mountain folds of the pleats that were locked to reveal a hexagon platform, slightly bent (Figure 2.2.4).

Extend the corners of the hexagonal platform with mountain folds to begin forming the pleats for the eventual hex twist. Lay those pleats flat CCW and twist the platform as you learned in Section 1.11 with the hex twist (Figure 2.2.5).

2.3 Triangle Twist Tessellation

Now, let's fold your first origami tessellation! You will start with the triangle twist tessellation. This is often an introductory tessellation. Because of this, you may see it called different names, such as *tiled hexagons* [2] (referring to the hexagonal floors it creates on the reverse) or the 3.3.3.3.3.3 tessellation (referring to the circles of six three-sided shapes around each floor).

This one uses a repetition of triangle twists, which you'll create by continuously splitting pleats at specific points. Begin with a CCW triangle twist in the center of the paper and split the pleat four grid lines away from the center of the triangle platform. Use the last two photos of Figure 2.3.1 as a reference of how to count grid lines between the triangle twist and the split point.

Reorient one of the pleats of the split so that it forms a second triangle twist. Now you have two triangle twists on the paper. Though there are two grid spaces between the closer corners of these twists, it is more accurate to consider their spacing to be four, as the unfolded distance between the centers of the triangles is four grid lines.

You may notice that the second twist naturally wants to rotate CW in order to lie flat, opposite of its neighbor. This will be a common theme throughout twist patterns. Split another pleat coming from the second twist, and you'll find the third twist also wants to rotate CCW, just like the first one (Figure 2.3.2).

Continue splitting in a circle, and you'll find that the fourth split generates a pleat that intersects with another already on the paper. Lock these pleats so that you can twist and flatten the fourth triangle. You will resolve the lock in the next step.

Hold on opposite sides of the lock as shown in the locking/unlocking skill diagram of Section 2.2. Pull the lock apart into a bar unlock with a width of four. This automatically creates the fifth and sixth splits; it's typically easier to do them both at the same time (Figure 2.3.3).

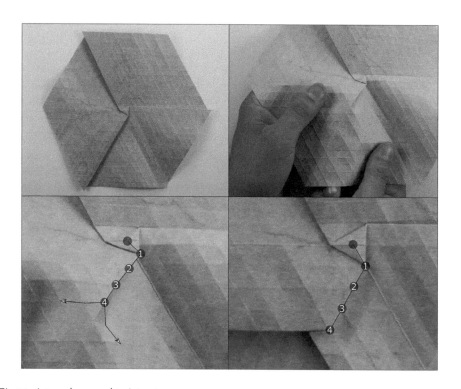

Figure 2.3.1 First twist and second twist setup.

Figure 2.3.2 Second, third, and fourth twist and first lock.

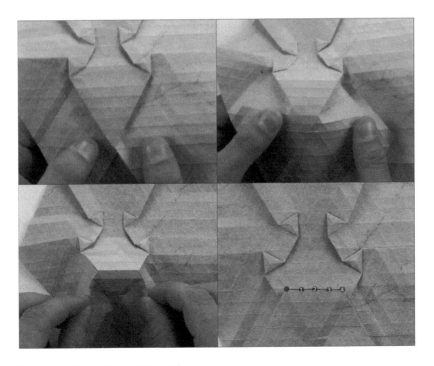

Figure 2.3.3 Unlocking into a bar with a width of four.

Figure 2.3.4 First, second, and third clusters.

Figure 2.3.5 LEFT: Finished triangle twist tessellation front. RIGHT: Finished triangle twist tessellation reverse.

Figure 2.3.6 TOP ROW: Spacing of two. BOTTOM ROW: Spacing of three.

I call a full circle of twists done in this manner a *cluster*. It is useful to work in clusters because it helps to keep your pattern organized. Continue the process of splitting and reorienting, until you run out of paper, locking and unlocking as necessary, and creating more clusters (Figure 2.3.4).

Figure 2.3.5 shows the finished triangle twist tessellation.

We chose a spacing of four at the beginning, but different choices of spacing lead to different densities of the patterning. Shown in Figure 2.3.6 are spacings of two and three, respectively.

2.4 Hexagon and Triangle Twist Tessellation

We used triangles to create the triangle twist tessellation; the tiling of triangles is a common arrangement for folders. Another common arrangement is to tile hexagons and triangles. Like the triangle twist tessellation, you will see this called different names, such as the 3.6.3.6 tessellation [2] due to there being two hexagons (six-sided shapes) and triangles (three-sided shapes) in a cluster around each floor. Imagine a line drawing of the pleats approaching a hex twist (in Figures 2.4.1–2.4.4, I mark the mountain folds only).

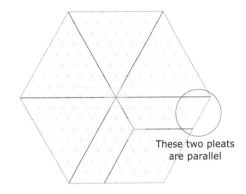

Figure 2.4.2 Triangle twist outline.

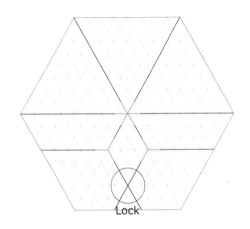

Figure 2.4.3 First lock outline.

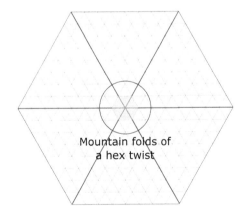

Figure 2.4.1 Hex twist outline.

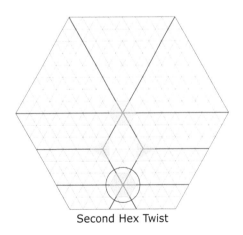

Figure 2.4.4 Second hex twist outline.

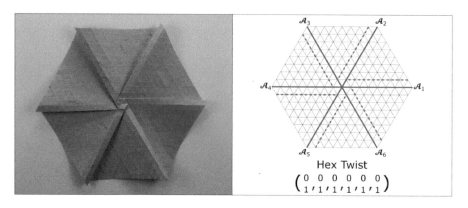

Figure 2.4.5 First hex twist.

Each line represents a simple pleat converging to form the hex twist.

If you split one of those pleats, the children pleats will run parallel to the two neighboring pleats.

If you split an adjacent parent pleat, its children pleats will intersect other already-formed pleats, forming a new lock.

You will turn that lock into a hex twist, resulting in the outline created in Figure 2.4.4. Knowing this, let's get to the folding!

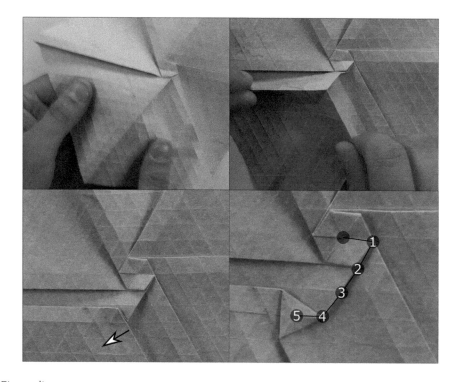

Figure 2.4.6 First split.

Figure 2.4.7 First row of triangle twists.

Begin with a CCW hex twist in the center of the grid.

Split one pleat coming from the hex twist. Counting from the center of the hexagon to the point of the split, make the split five grid lines away from the center of the hexagon.

Turn that split into a triangle twist (Figure 2.4.6).

Repeat the splitting with each pleat coming from the hex twist (Figure 2.4.7). This creates six distinct locks, which you will resolve in the next step.

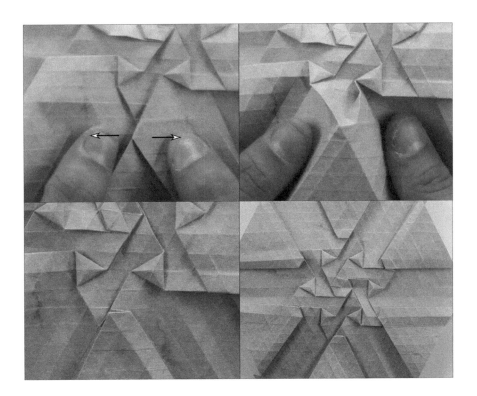

Figure 2.4.8 Resolving the locks.

Figure 2.4.9 Unlocking into a hex twist.

Pull apart one of the locks and flatten it into a bar with a width of one. Do this to all six locks (Figure 2.4.8).

Next, turn those bars into hex twists. Pull apart a bar as shown in Figure 2.4.9. Identify the grid intersection where the two folded pleats' mountain folds intersect and pinch the grid lines that also intersect that point.

Continue pinching the pleats that will create the hex twist and lay them flat CCW. Twist and flatten the second hex twist (Figure 2.4.10). Notice that pleats coming off this second hex twist lock with pleats you made when creating locks during the first round of triangle twists. You will resolve those locks in future steps.

Figure 2.4.10 Second hex twist.

Figure 2.4.11 Creating the third hex twist.

Figure 2.4.12 Triangle twists between the hex twists.

Figure 2.4.13 LEFT: Finished triangle and hexagon twist tessellation front. RIGHT: Finished triangle and hexagon twist tessellation reverse.

Figure 2.4.14 Spacing of four.

One at a time, resolve each of these new locks into a hex twist (Figure 2.4.11).

You'll notice that, as you form the hex and triangle twists, the new hex twists will begin to resemble the center hex twist and will be partially surrounded by triangle twists. You'll want to complete the pattern so each hex twist is entirely surrounded by six triangle twists.

Tile the rest of the paper this way. Figure 2.4.13 shows the finished tessellation.

If the triangle twists were only four spaces away from the hex twists, Figure 2.4.14 shows what the final tessellation would look like.

2.5 Tessellation Basics (without Folding)

Now that you've folded two origami tessellation patterns, it will help to have some background in the underlying geometry of tessellations. From there, you can put what you folded in the last two sections into more context.

In geometric terms, a tessellation is a two-dimensional tiling of shapes without holes or overlaps. The shapes can be *polygons* (closed shapes with straight edges and no holes) or shapes with curves.

Figure 2.5.1 LEFT: Arbitrary tessellation. RIGHT: Arbitrary tessellation with curves.

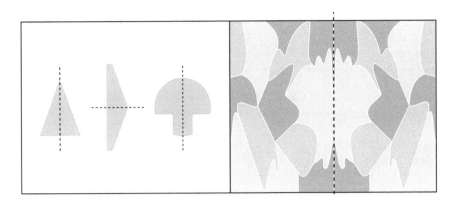

Figure 2.5.2 LEFT: Bilateral symmetry. RIGHT: Arbitrary tessellation with bilateral symmetry.

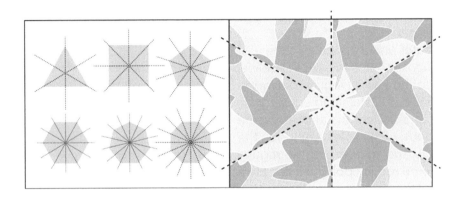

Figure 2.5.3 LEFT: Various radial symmetries. RIGHT: Arbitrary tessellation with hexagonal symmetry.

A common feature of tessellation patterns and individual tessellation tiles is *symmetry*. Figure 2.5.2 show tiles and a pattern with *bilateral symmetry*, meaning that the two halves can be flipped or folded onto each other and map point for point onto the opposite half.

The shapes in Figure 2.5.3 display *radial symmetry*, which means the tile/pattern between the dotted lines maps onto each similar section by rotating about the center. More specifically, a tile/pattern can display three-fold, four-fold, five-fold, or greater symmetry. The tessellation on the right of the same figure displays six-fold symmetry.

A *periodic tessellation* is one with a consistent pattern of the same arrangement of

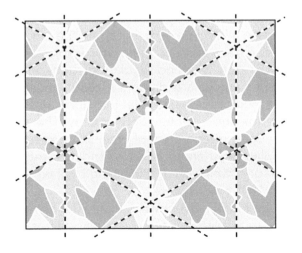

Figure 2.5.4 Arbitrary tessellation with periodic symmetry.

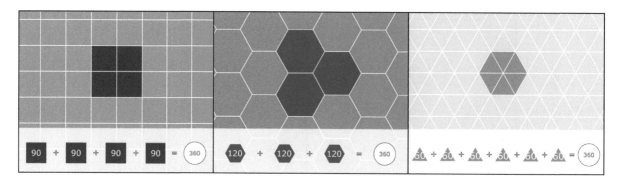

Figure 2.5.5 Regular tessellations. LEFT: 4^4 tiling. MIDDLE: 6^3 tiling. RIGHT: 3^6 tiling.

tiles throughout. Figure 2.5.4 uses a similar tile in six-fold symmetry copied throughout and is considered to be periodic.

To attain a greater understanding of the vast world of tessellations, it helps to have a baseline, so let's go back to the basics: regular polygons. *Regular polygons* are *equilateral* and *equiangular*, i.e. with the same side lengths and the same angles throughout. Examples include the equilateral triangle, the square, and the regular hexagon. Only some regular polygons can tile a flat plane indefinitely. In most tessellations, the corners of several shapes will come together at a single point, called a vertex. To create a flat tessellation with no holes or overlaps, all of the angles around a single vertex must add to 360° to make a full circle. Any more, and there will be overlaps; any less, and there will be holes.

Your goal is to find regular polygons whose interior angles are divisors of 360°, meaning that multiple copies of that polygon can share a vertex and complete 360° exactly. The ones that work independent of other shapes are the equilateral triangle, square, and the regular hexagon, which have interior angles of 60°, 90°, and 120°, respectively.

The four tilings shown in Figure 2.5.5 are the only *regular tessellations*, meaning tilings that are composed of the same, regular polygon with the same size. Geometers name regular tessellations based on the number of sides of the shapes around any given vertex. The triangle tessellation has six three-sided shapes (triangles) around each vertex, so it is commonly called the 3.3.3.3.3.3 tessellation, or 3^6. The square tessellation has four four-sided regular shapes (squares) and is thus called a 4.4.4.4, or 4^4 tessellation. The hexagonal tessellation has three six-sided regular polygons and is thus called a 6.6.6, or 6^3 tessellation.

By mixing regular polygons with similar side lengths, you can discover that the interior angles of two triangles and two hexagons also sum to 360°; when connected at their vertices, they create the top-left tessellation in Figure 2.5.6. Although I use two different polygons, each vertex in this tessellation is identical, allowing for rotation: each connects a triangle, a hexagon, a triangle, and another hexagon. So, this tessellation is called a 3.6.3.6 pattern. This pattern will be used as often as the 3^6 and the 6^3 patterns throughout this book. A tessellation that uses a mixture of different regular polygons—like the 3.6.3.6 pattern—but has the same vertex throughout is called a *semiregular tessellation*. Together the regular and semiregular tessellations are called *Archimedean tilings*. There are eight semiregular tilings and three regular tilings, leading to eleven Archimedean tilings total. The semiregular tilings can be seen in Figure 2.5.6.

Tessellation study becomes particularly interesting when you consider *dual tessellations*,

Figure 2.5.6 Semiregular tessellations. TOP ROW: 3.6.3.6 and 3.3.3.3.6 tiling. SECOND ROW: 3.4.6.4 and 3.3.3.4.4 tiling. THIRD ROW: 3.3.4.3.4 and 4.8.8 tiling. FOURTH ROW: 4.6.12 and 3.12.12 tiling.

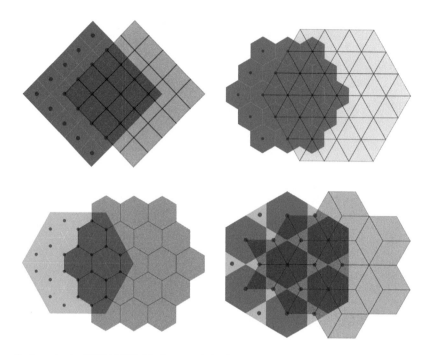

Figure 2.5.7 Tessellation duals. TOP LEFT: 4^4 tiling and dual. TOP RIGHT: 6^3 tiling and dual. BOTTOM LEFT 3^6 tiling and dual. BOTTOM RIGHT: 3.6.3.6 tiling and dual.

created by connecting the centers of a tessellation's tiles. The resulting tessellation is considered the dual of the initial tessellation. Figure 2.5.7 show the duals for the 3^6 tiling, 4^4 tiling, 6^3 tiling, and 3.6.3.6 tiling.

There are three observations I'd like to make about these figures. The first is that the dual to a 4^4 pattern is another 4^4, making it a *self-dual*. The second is that the dual of a 6^3 tiling is the 3^6 pattern, and vice versa. In fact, you'll find that the dual of any dual is initial tessellation itself. The third thing to note is that the dual of the 3.6.3.6 tessellation results in a tessellation composed of a nonequiangular shape, a rhombus with two 60° angles and two 120° angles. This resultant tessellation also has two distinct vertex assortments, one with six rhombi coming together, and one with three coming together (Figure 2.5.8).

This tiling is not Archimedean, but it is of importance to folding pleat patterns, given its

strong relationship to the 3.6.3.6 tessellation. It is called the *rhombille tiling*. Since this tiling is not composed of regular polygons, it does not receive a numbered vertex designation.

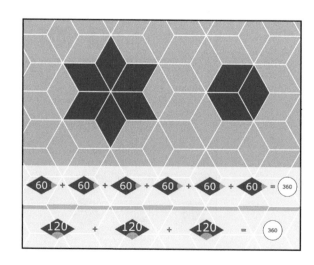

Figure 2.5.8 Rhombille tessellation.

2.6 Applying Tessellation Knowledge to Folding

If tessellations are tilings of a plane without holes and do not use cuts, then couldn't every flat foldable origami piece potentially be considered a tessellation? The tiles do not have to be regular for a design to be considered a tessellation. Further thought may lead you to notice that the tiles of a tessellation aren't allowed to overlap;

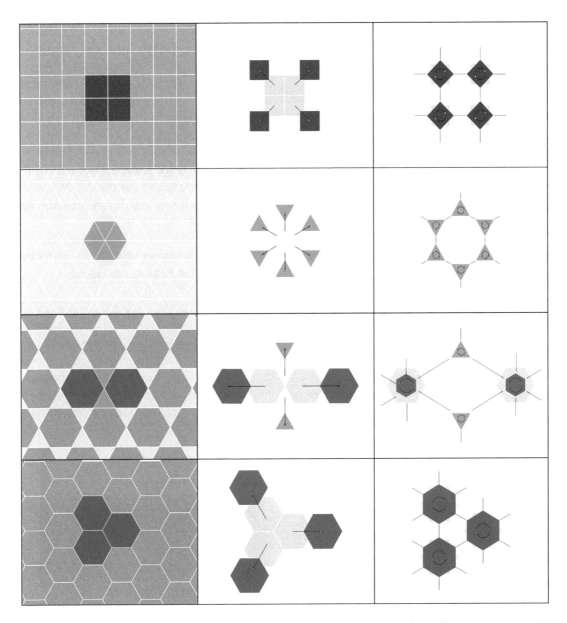

Figure 2.6.1 Exploding a tiling to represent an origami tessellation. TOP ROW: 4^4 tessellation. SECOND ROW: 3^6 tessellation. THIRD ROW: 3.6.3.6 tessellation. FOURTH ROW: 6^3 tessellation.

as a pleat is an overlap, origami breaks that rule from the outset. What if you consider just the twists (as many folders do and you did with the triangle and hexagon tessellation in Section 2.4) to be the "tiles" that create the tessellation? If so, this method of describing the tessellation has the twists be separate from each other, connected by pleats; the pleats encircle the floors, which are analogous to holes in the tiling. The existence of holes disqualifies the model from being a tessellation. The only things that are really tiled without holes and without overlaps are the floors when viewed from the reverse of the paper, though even this is only true in special cases. The relevance of tessellation study is the fact that much of the geometry for origami tessellations is strikingly similar to that of "true" tessellations. For example, a triangle twist tessellation naturally tends to have six triangles in a cluster around a hexagonal floor, simply due to the geometry. For this reason, the notion of a *tessellation* in the mathematical sense is different from the notion of an *origami tessellation*, referring to tilings of origami twists on a folded plane.

Figure 2.6.1 shows a process for turning an Archimedean tiling into an origami tessellation. To translate into a foldable tessellation, it helps to simplify the tiling and focus on a single vertex of polygons. In the left column of the figure, the polygons of the tilings represent twists, which will be connected by pleats in the right column. Take the polygons that share a vertex and spread them out radially the same distance from the connecting vertex. The further you spread these polygons, the less dense the folded tessellation will be. Separating the polygons in this manner is called *exploding* the tessellation. After exploding the tessellation, rotate the polygons until the corners point toward each other, such that all points of the spread-out polygons can be connected with nonintersecting lines. You also have to shrink the hexagons in the 3.6.3.6 pattern to align with the projected size of the triangles, forming parallel pleats. Figure 2.6.1 shows the procedure for the 4^4, 3^6, 3.6.3.6, and 6^3 tilings. With the last figure in each row, you have a more accurate representation of what an origami tessellation will look like. You can use what you've learned about tilings to broaden your repertoire of origami designs.

2.7 Triangle Weave Pattern

The triangle twist is composed of three intersecting pleats. Likewise, so is the triangle spread twist. The hex twist is composed of six intersecting pleats. Likewise, so is the hex spread. The spreads are analogous to their regular

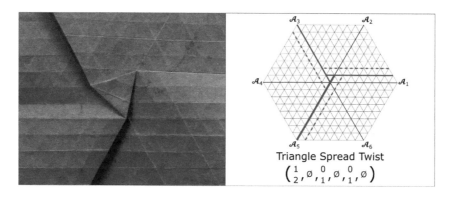

Figure 2.7.1 Starting triangle spread and pleat schematic.

Figure 2.7.2 Setting up the second twist.

twists, except that their pleats are drifted from the center. This creates a spread, an opening on the reverse of the twist, which produces an interesting weave effect when tiled.

Start with a triangle spread twist in the middle of the paper. Notice that there's no way to place the twist at the exact center of the grid.

As with the triangle twist tessellation you learned in Section 2.3, split one of the pleats four spaces away from the spread twist. Once split, drift the pleat held in the right photo of Figure 2.7.2 one space CW.

Figure 2.7.3 shows the counting. With a spread, I find it more useful to count from the corner of the platform when the spread is standing, since the "center" of the triangle is not on a grid line.

Flatten the second twist (Figure 2.7.4). Keep creating triangle spreads, alternating between CW and CCW rotations and a corresponding pleat drift for each. Notice that when you form the fourth twist, there is a lock that needs to be resolved (Figure 2.7.5).

As before, hold on the larger angles of the lock and pull them apart. Like with the triangle twist tessellation, it is helpful to create the standing form of the remaining two triangle spreads at the same time. Do this with simultaneous splits and drifts, or just pinch at the appropriate grid lines to create a bar with a width of four. Use the counting in the right photo of Figure 2.7.6 as a guide for this.

Flatten the remaining twists in the cluster. Keep creating twists on the paper using the same method throughout the paper (Figure 2.7.7).

Figure 2.7.8 shows the finished triangle spread tessellation. The negative space on the reverse side of the paper really stands out in the full pattern.

Just as in the triangle twist tessellation, different spacings modify the density of the tessellation. Figure 2.7.9 shows examples with a spacing of two and three, respectively.

Figure 2.7.3 Counting a spacing of four.

Figure 2.7.4 The second twist.

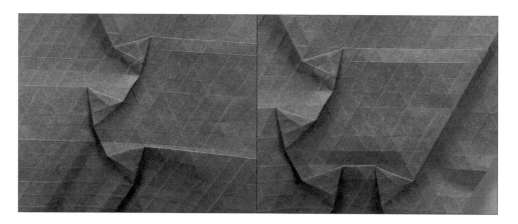

Figure 2.7.5 Third and fourth twists and first lock.

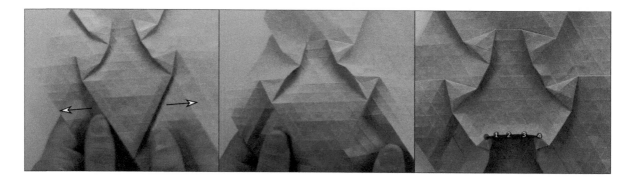

Figure 2.7.6 Unlocking into a bar with a width of four.

Figure 2.7.7 First and second cluster.

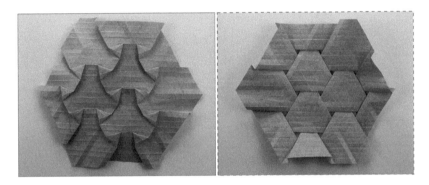

Figure 2.7.8 LEFT: Finished triangle weave tessellation front. RIGHT: Finished triangle weave tessellation reverse.

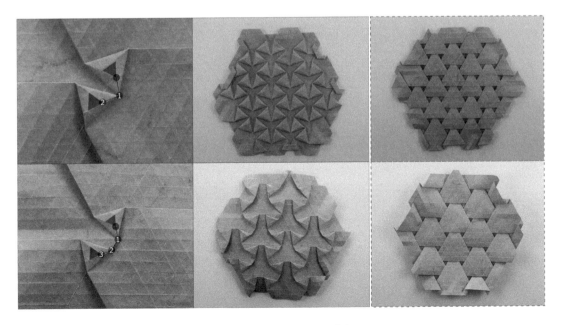

Figure 2.7.9 TOP ROW: Spacing of two. BOTTOM ROW: Spacing of three.

2.8 The Hex Weave Pattern

You've just seen that tessellations involving the triangle twist and triangle spread twist follow similar patterns. This is not to say these twists are interchangeable, but their geometry is analogous. Similarly, so is the geometry between the hex twist and the hex spread twist. The tessellation you'll learn in this section uses a combination of hex spreads

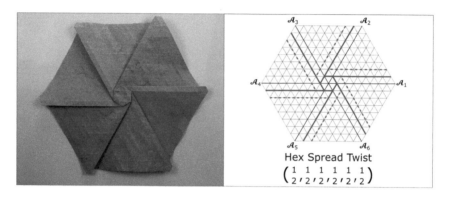

Figure 2.8.1 Starting hex spread and pleat schematic.

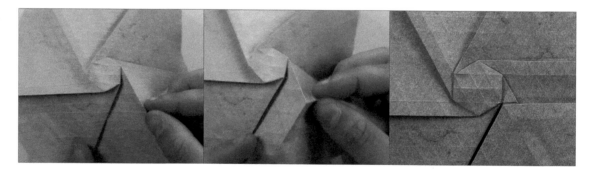

Figure 2.8.2 First split.

Figure 2.8.3 First triangle twist and counting a spacing of four.

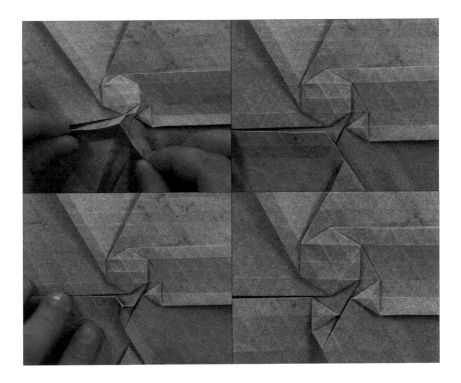

Figure 2.8.4 First lock.

and triangle twists. You could just as well use hex spreads and triangle spreads, but the contrast between the intense negative space of the hex spread and the hiddenness of the triangle twist on the paper's reverse enhances the "weave" effect in a way that makes for a more striking pattern. For this reason, this tessellation is sometimes called a hex weave or open-back hex pattern [2] instead of being labeled as a 3.6.3.6 pattern.

Begin with a hex spread in the middle of the paper.

Split one of the pleats coming from the hex spread so that the children pleats, when oriented toward the hex spread, lie aligned with their neighbor pleat as shown in Figure 2.8.2.

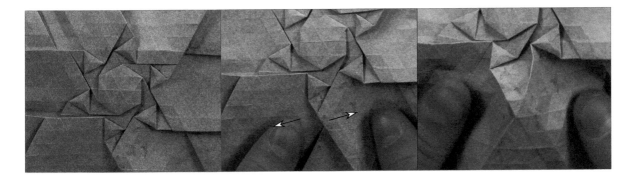

Figure 2.8.5 First unlock.

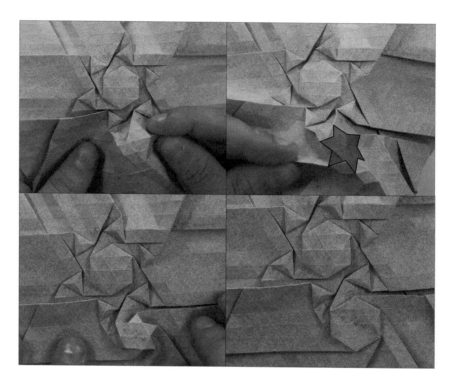

Figure 2.8.6 Second hex spread twist.

Orient the children pleats of the split so they rotate in the same direction and create a CW triangle twist. The second photo of Figure 2.8.3 shows counting the spaces from between the hex spread and the triangle twist center. Just as with the triangle spread, it is easier to count from the corner of the hex spread's interior hexagon (the platform when the hex spread is standing). In Figure 2.8.3, the triangle twist is four spaces away from the hex spread.

Split the next pleat and notice that one of the children of that split creates a lock with a previous pleat. You'll resolve that in a later step. For now, use the split to create a triangle twist, and repeat with all of the parent pleats coming off the central hex spread (Figure 2.8.4).

Now you'll resolve the locks. Choose a pair of locked pleats and unlock them as shown in Figure 2.8.5.

Pinch the pleats to form a new hex spread where the lock was. Use the highlighted star in the second photo of Figure 2.8.6 as a guide to locate the hex spread's platform. Pinch these new pleats all the way to the edge of the paper, one at a time. Then collapse all at once to form the standing hex spread, and collapse.

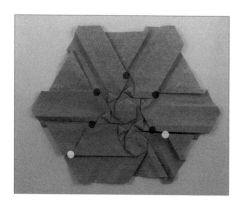

Figure 2.8.7 Planning the next locks.

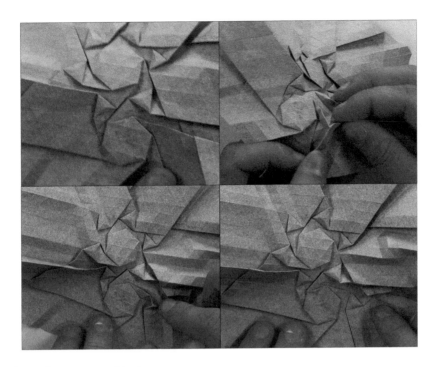

Figure 2.8.8 Resolving the next set of locks.

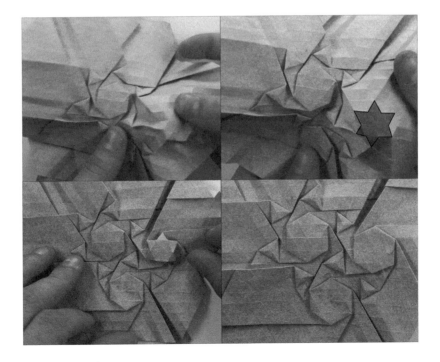

Figure 2.8.9 The third hex spread.

Figure 2.8.10 Standing forms of the remaining hex spreads.

Notice the several locks, marked in circles in Figure 2.8.7. You can ignore the ones marked white, since these will automatically be resolved as the result of resolving another, more interior lock. In Figure 2.8.8, you will resolve the next CCW lock, and then continue resolving the twists surrounding the center.

Your next goal is to create a cluster of three hex spreads. Before you can do that, you need to introduce triangle twists between them, to replicate the circle of triangle twists surrounding the central hex spread. Notice the two triangle twists at the top of the hex spread you last made (third photo of Figure 2.8.8). Locate the next CW pleat coming from this second hex spread and split it three spaces away, as you've done before. Use that split to create a new triangle twist (fourth photo of Figure 2.8.8).

Now you have a lock of three pleats, each offset from what will be the center of the next hex spread. Open the lock (first photo of Figure 2.8.9), find the star, and create the pleats that go into the hex spread. As before, extend these new pleats all the way to the edge of the paper. Continue the pattern of unlocking, finding the star, and creating the pleats for the next spread.

It can be helpful and efficient to set the pleats before flattening everything. In Figure 2.8.10,

Figure 2.8.11 TOP ROW: Second row finished. BOTTOM LEFT: Finished hex weave front. BOTTOM RIGHT: Finished hex weave reverse.

Figure 2.8.12 Spacing of three.

I keep the hex spread and triangle twists in standing form until I have everything in place.

Figure 2.8.11 shows the finished hex weave. Figure 2.8.12 shows an alternate spacing of three. With this spacing, the folder has to choose which pleat lies atop the other, and it gets a little crowded for beginning folders. However, if you can master this; it really displays the weave effect.

2.9 Hexagonal Failed Cluster

As discussed in Section 2.5, there are three regular tilings: the 3^6, the 4^4, and the 6^3. The 4^4 is folded on a square grid. The 3^6 tiling can be folded as the triangle tessellation. What would the 6^3 mathematical tessellation look like when translated to an origami form? As it turns out, this tiling carries a complication, which becomes apparent when you try to fold it. To truly understand this complication, it is helpful to fold the pattern, so I encourage you to follow along nonetheless.

Start with a hex twist in the center of the paper rotating CCW.

You're going to create the second hex twist seven grid lines away from the center of the first hex twist with a CW rotation. To do this, split one of the pleats seven spaces away from the center of the first twist (Figure 2.9.2). This creates three of the six pleats that create the hex twist. You need three more pleats. Form them from the edges inward. Two of them will lock with existing pleats, which you'll resolve in a future step.

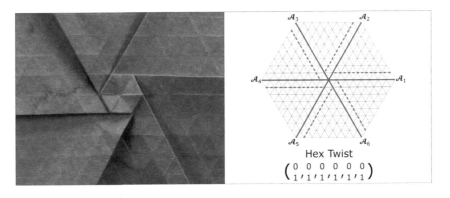

Figure 2.9.1 Starting hex twist and pleat schematic.

Figure 2.9.2 Splitting and setting up the mountain folds of the next hex twist.

Figure 2.9.3 shows what the second twist with CW rotation looks like when folded flat.

The next twist will occur at one of the pleat locks formed by the second twist. Unlock them as usual (Figure 2.9.4).

Use that lock to form the hexagon (Figure 2.9.5). Unfortunately, you will encounter an issue finishing the third hex twist: the first twist rotates CCW, and the second rotates CW, so the third one should have… which rotation? In fact, it doesn't have a valid rotation. You can resolve the problem by forcing the pleats into different orientations as shown in the left image of Figure 2.9.5, but this is not ideal.

Figure 2.9.3 Second twist and counting a spacing of seven.

Figure 2.9.4 Third twist setup.

Figure 2.9.5 Trying to lay the third twist flat.

Figure 2.9.6 LEFT: "Finished" hex twist tessellation front. RIGHT: "Finished" hex twist tessellation reverse.

Figure 2.9.6 shows what the tesselation looks like when all of the hexagons except the center are rotating CCW, despite your loose "rule" that every twist must rotate opposite of its neighbors. When arranged this way, the tessellation's pleats do not lie flat on the table and change orientation halfway between molecules. Using the techniques you know so far, this is as resolved and as flat as you can get the twists. In the following sections, you'll learn alternate workarounds.

2.10 Modifications

The tessellations you've explored so far have been focused on flat-foldable twists. The next sections will explore different ways of modifying twists to account for issues—like the one brought up in Section 2.9—or to create decorations for the tessellations that "work." First, let's review the basic elements of the six simple twists you already know: the pleat intersection, the standing form, the flattened form, and the reverse.

Using Figures 2.10.1–2.10.6, take a moment with each twist to really internalize what these elements look like. As you grow to recognize them, you'll start to see them appearing often as your own patterns develop. Note that the standing form of the hex twist introduces creases that are not on the grid. For this reason, I generally skip the standing "step" in folding it, and this is one of the reasons folding a hex twist is so difficult for beginners.

Later sections will explore modifications of these molecules, but for now I'll focus on a special type of modification: backtwisting, which opens the door to sinking and expanding a molecule or flattening the pleats. Figure 2.10.7 shows examples of backtwisting, sinking, expanding and flattening the pleats of the triangle, hex, and arrow twists. Should you wish to explore these forms more fully, these modifications are diagrammed in the following sections:

- Backtwisting: Section 2.11
- Flattened Pleats: Section 2.13
- Twist Sinking: Section 2.18
- Twist Expansion: Section 2.19

As you backtwist, sink, drift, reorient, or generally change certain features of molecules, you may find particular changes you enjoy using more than others. The button, kite, and nub molecules (Figure 2.10.8) are some of my favorites, and they will be used in tessellations in later sections as well, the latter two of which are results of reorienting pleats of a rhombic twist. Note that they are

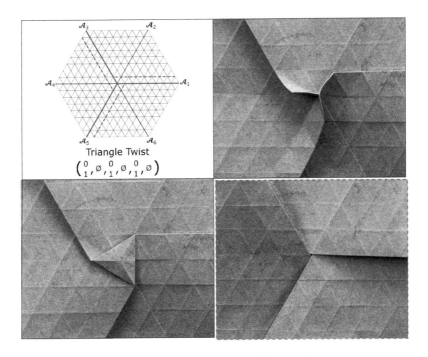

Figure 2.10.1 Triangle twist. TOP LEFT: Pleat schematic. TOP RIGHT: Standing form. LOWER LEFT: Twist. LOWER RIGHT: Reverse.

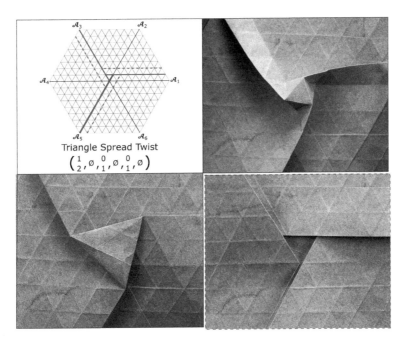

Figure 2.10.2 Triangle spread twist. TOP LEFT: Pleat schematic. TOP RIGHT: Standing form. LOWER LEFT: Twist. LOWER RIGHT: Reverse.

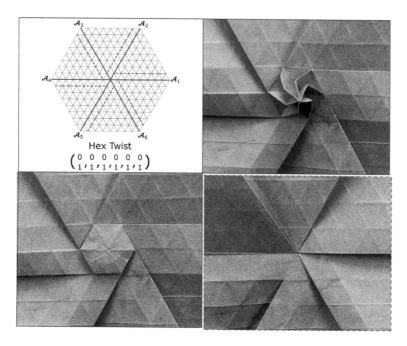

Figure 2.10.3 Hex twist. TOP LEFT: Pleat schematic. TOP RIGHT: Standing form. LOWER LEFT: Twist. LOWER RIGHT: Reverse.

Figure 2.10.4 Hex spread twist. TOP LEFT: Pleat schematic. TOP RIGHT: Standing form. LOWER LEFT: Twist. LOWER RIGHT: Reverse.

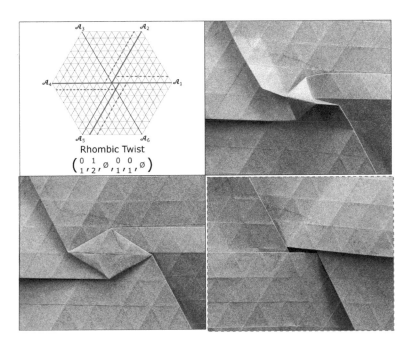

Figure 2.10.5 Rhombic twist. TOP LEFT: Pleat schematic. TOP RIGHT: Standing form. LOWER LEFT: Twist. LOWER RIGHT: Reverse.

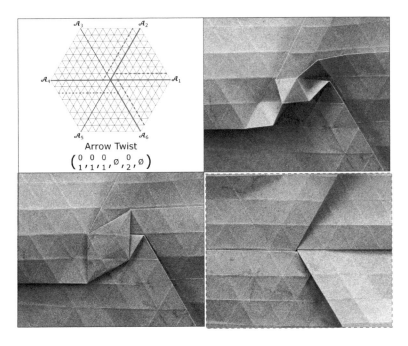

Figure 2.10.6 Arrow twist. TOP LEFT: Pleat schematic. TOP RIGHT: Standing form. LOWER LEFT: Twist. LOWER RIGHT: Reverse.

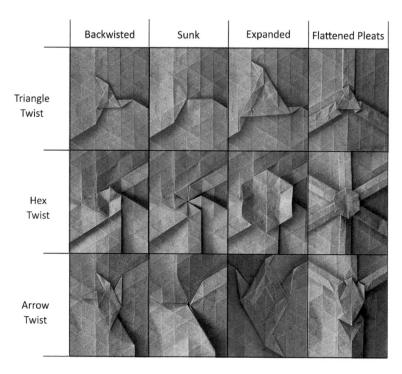

Figure 2.10.7 Modifications of the six simple twists: Backtwisted, sunk, expanded, and with flattened pleats. TOP ROW: Triangle twist. SECOND ROW: Hex twist. THIRD ROW: Arrow twist.

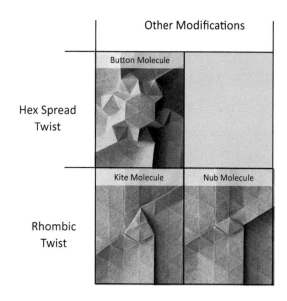

Figure 2.10.8 TOP LEFT: Button molecule modification of the hex twist. BOTTOM ROW: Kite and nub molecule modifications of the rhombic twist.

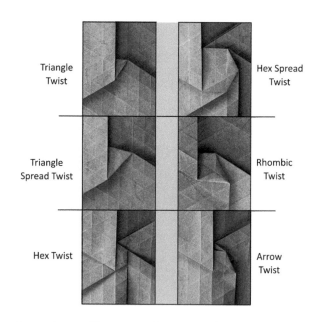

Figure 2.10.9 The six simple twists with one pleat inverted.

not called twists since they do not rotate in the same way as the forms you've learned before, so the more generic term "molecule" is used. These variant diagrams are discussed in the following sections:

* Kite Molecule: Section 2.17
* Nub Molecule: Section 2.17
* Button Molecule: Section 2.23

Lastly, Figure 2.10.9 shows examples of *inverting* one pleat of each of the six simple twists. *Pleat inversion* will be discussed in Section 2.34 and opens up an entirely new way of interpreting molecules.

2.11 Backtwisting

Backtwisting can be applied to the triangle twist, the hex twist, and the arrow twist. After

they are backtwisted, you can call them the *triangle backtwist*, *hex backtwist*, and *arrow backtwist*, respectively.

To backtwist a triangle, begin with a triangle twist and pinch into standing position. Pinch the extension from the triangle corner down as shown in Figure 2.11.1. Flatten the triangle

Figure 2.11.2 Reorienting a pleat after a backtwist.

Figure 2.11.1 Triangle backtwist.

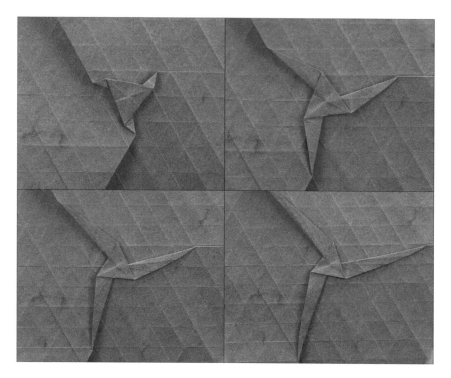

Figure 2.11.3 Different degrees of triangle backtwist.

in the opposite rotation of the pleats. Note that new creases will be created in the process. Backtwisting frees the pleats to move separately from the rest of the molecule.

The photo in Figure 2.11.2 shows a triangle backtwist with one pleat reoriented.

Figure 2.11.4 Side view of a regular hex twist.

There is not really an upper limit to the angle of backtwisting you can perform. Figure 2.11.3 shows several versions of a triangle backtwist with different angles of rotation. I usually just rotate until the corner hits the next grid intersection, which is a 60° rotation. This is enough to release the pleats and keeps everything well structured. In this text when I talk about a "backtwist," the 60° rotation is what I will mean, but you should keep the alternate options in mind.

Consider a hex twist as viewed from the side (Figure 2.11.4). Notice how the mountain fold of each pleat connects directly into the corner of the hexagon platform. This will change once the backtwisting has been completed.

To create this backtwist, there's no need to stand the molecule like with the triangle. Lift up three corners of the hexagon and twist them in the opposite rotation from the pleats. Then lay down flat as shown in Figure 2.11.5.

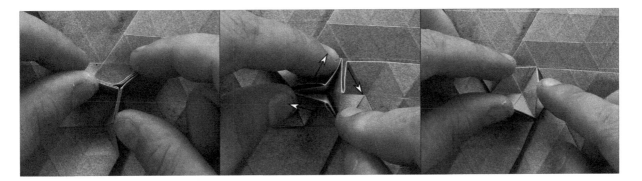

Figure 2.11.5 Backtwisting the hexagon platform.

Figure 2.11.6 Side view of a hexagonal backtwist.

When you've finished, the top view of the backtwist will look almost identical to the typical hex twist, but the side view (Figure 2.11.6) reveals that the mountain folds of the pleats no longer connect directly to the corners of the hexagon. As with the triangle backtwist, this allows the pleats to move separately from the rest of the hex twist.

The arrow backtwist has a similar format to the hex and triangle backtwists. Stand the molecule and rotate the twist in the opposite direction of the pleats. Figure 2.11.7 shows what the arrow backtwist looks like when finished.

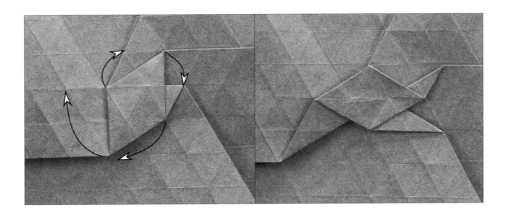

Figure 2.11.7 Backtwisting the arrow twist.

2.12 Twist Handedness and Pleat Symmetry

Once you folded a cluster of three hex twists in the "failed" hex twist tessellation, it was apparent why the tiling didn't lie flat as neatly as regular twist patterns. The twists had a particular rotation, and the pattern was doomed to clash because of the odd number of molecules in each cluster. When there is an even number of molecules in a cluster (such as six triangles or two triangles and two hexagons), the twist rotations can neatly alternate between CCW and CW for each molecule. I like to think of the twists as gears that rotate opposite to each adjacent gear.

When you fold simple twists, the pleats have a particular orientation, depending on whether you collapse the pleat's standing form in a CW or CCW direction. Reorienting changes the side the valley fold is on. You can visualize this orientation choice by viewing a cross-section of a parallel pleat, which I call the *pleat profile*; the profiles are shown on the bottom of each drawing in Figure 2.12.1 and correspond to the creases marked above them.

The 6.6.6 cluster shown in the last diagram doesn't "work" if the pleats have to lie flat. It's possible to have a perfectly reasonable tessellation with the twists all rotating in the same direction, but the pleat will be upright and skewed, as you can see in the photo of Figure 2.12.2.

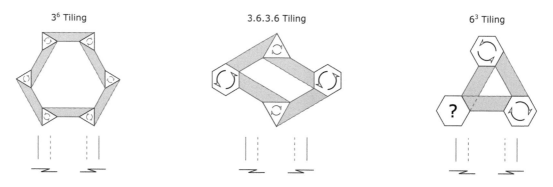

Figure 2.12.1 Twist handedness and tilings.

Figure 2.12.2 6.6.6 failed cluster.

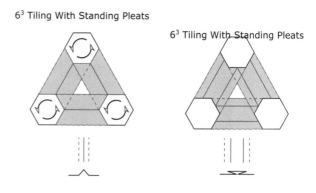

Figure 2.12.3 Symmetric forms of pleats. LEFT: Standing form. RIGHT: Flattened form.

The pleats connecting the first cluster of hexagons are in standing form, if a bit distorted. The pleat profile would look like those in Figure 2.12.2. In this case, both the possible valley folds are partially folded.

This pleat has bilateral symmetry, and therefore is allowed to connect two adjacent twists. The hex twists are even free to have a defined rotation (CCW, in this case), relatively independent of the pleat orientation. However, when you try to lay the pleats flat and maintain the hexagon rotation, the pleats would resist and tear somewhere in the middle. To truly get a flat-folded connection between the twists, we need a symmetric pleat that is also able to fold flat without tearing.

The figure on the right of Figure 2.12.3 has no handedness, which can be seen because its profile is also symmetric, and thus there is no preferred rotation for the twists. You can use this to your advantage when folding a 6^3 design.

2.13 Pleat Flattening

I will call the pleat on the right in Figure 2.12.3 a *flattened* pleat, and this section will present its folding process. The flattened pleat is your first composite pleat—a pleat that has more than a single pairing of mountain and valley folds. It can be helpful to think of it as two separate simple pleats oriented away from each other.

Sometimes you may only have a single-wide pleat that you decide you want to flatten. Rather than unfold and bring the gridding from 32nds to 64ths (for example), it can be helpful to know

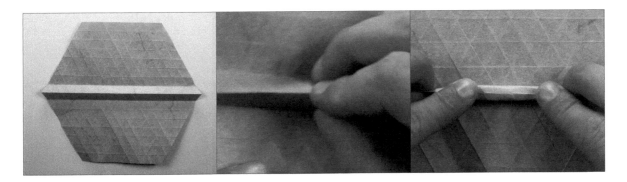

Figure 2.13.1 Standing pleat and starting the flattening.

Figure 2.13.2 Flattening the pleat.

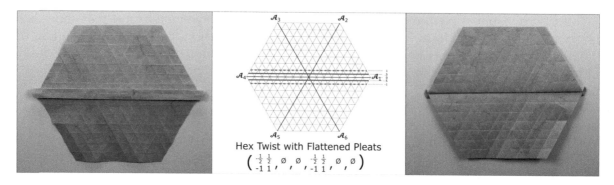

Hex Twist with Flattened Pleats
$$\begin{pmatrix} \frac{1}{2} \frac{1}{2} & \varnothing & \varnothing & \frac{1}{2} \frac{1}{2} & \varnothing & \varnothing \\ -1\,1 & & & -1\,1 & & \end{pmatrix}$$

Figure 2.13.3 LEFT: Single-wide pleat flattened front. MIDDLE: Pleat schematic for the single-wide pleat flattened. RIGHT: Single-wide pleat flattened reverse.

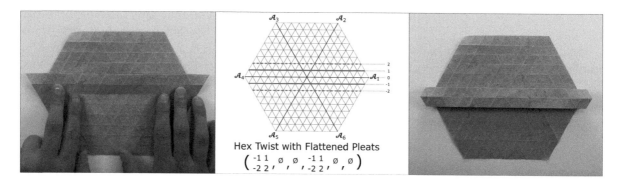

Hex Twist with Flattened Pleats
$$\begin{pmatrix} -1\,1 & \varnothing & \varnothing & -1\,1 & \varnothing & \varnothing \\ -2\,2 & & & -2\,2 & & \end{pmatrix}$$

Figure 2.13.4 LEFT: Double-wide simple pleat. MIDDLE: Pleat schematic for the double-wide pleat flattened. RIGHT: Double-wide pleat flattened reverse.

how to flatten the pleat, even with a less dense grid. Stand the pleat and pinch downward in the middle of the pleat, separating the two layers on either side. This will create new creases between grid lines (Figure 2.13.1).

Pressing down with your fingers, slowly slide your fingertips toward one side of the pleat until it hits the edge of the paper. Then, slide them toward the other edge of the paper (Figure 2.13.2).

When you finish, Figure 2.13.3 shows what your flattened pleat will look like. Notation can get a little unwieldy when talking about composite pleats, especially those describing spacing between grid lines. The notation in

Figure 2.13.3 uses fractional values because you flattened pleats at the smallest grid subdivision, resulting in creases between grid subdivisions.

You can simplify the notation—as well as the folding—by starting with a double-wide initial pleat so that the flattened pleats map onto the smallest subdivision, as in Figure 2.13.4.

Hex Twist with Flattened Pleats

When flattening a pleat connected to a twist, you need to backtwist first. In Figures 2.13.5–2.13.8, I use a hexagonal backtwist as an example. Begin with a hexagonal backtwist and reorient one of the pleats.

Figure 2.13.5 Hexagonal backtwist and reorienting one pleat.

Figure 2.13.6 Standing the pleat and flattening.

Figure 2.13.7 Two pleats flattened.

If the pleats around the hex backtwist are all oriented CCW, when you reorient one of the pleats, the next adjacent pleat in the CCW direction will be surrounded by pleats pointing away from it. This is the pleat you'll start flattening (Figure 2.13.5). Stand that pleat and flatten it to the edge of the paper. Then walk it in toward the hex backtwist, as far as it can go neatly; it will introduce new creases into the paper from the mountain folds of the pleat to the hexagon platform's corner (Figure 2.13.6).

Then flatten the hexagon and repeat with the next pleat around in a circle. For the second pleat (and every pleat thereafter) you will have to choose a layer ordering. I usually just put each pleat on top of the pleat CW of it (Figure 2.13.7).

Figure 2.13.8 shows the finished hex twist with flattened pleats.

Using the same technique used in Figure 2.13.4, you can start with double-wide pleats and flatten them, simplifying the notation and folding process (Figure 2.13.9).

Hex Twist with Partially Flattened Pleats

You may have noticed that, while you can flatten a pleat as far is it can go, there are other points where you might stop flattening earlier. I call not extending the pleat flattening as far as it can go *partial flattening*.

To partially flatten a pleat, begin with the same steps as the full flattening. Start with

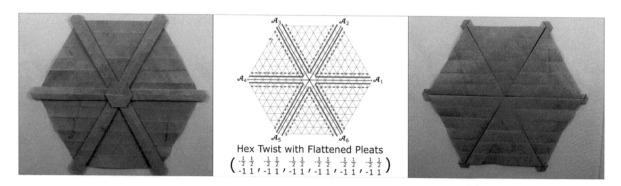

Hex Twist with Flattened Pleats
$$\begin{pmatrix} \frac{-1}{2}\frac{1}{2} & \frac{-1}{2}\frac{1}{2} & \frac{-1}{2}\frac{1}{2} & \frac{-1}{2}\frac{1}{2} & \frac{-1}{2}\frac{1}{2} & \frac{-1}{2}\frac{1}{2} \\ -1\ 1\ ' & -1\ 1\ ' & -1\ 1\ ' & -1\ 1\ ' & -1\ 1\ ' & -1\ 1 \end{pmatrix}$$

Figure 2.13.8 LEFT: Finished hex twist with flattened pleats front. MIDDLE: Schematic of hex twist with flattened pleats. RIGHT: Finished hex twist with flattened pleats reverse.

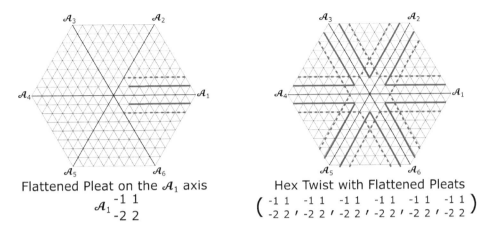

Flattened Pleat on the $\mathcal{A}_1$ axis

$$\mathcal{A}_1 \begin{matrix} -1 \ 1 \\ -2 \ 2 \end{matrix}$$

Hex Twist with Flattened Pleats

$$\left(\begin{matrix} -1 \ 1 \\ -2 \ 2 \end{matrix} , \begin{matrix} -1 \ 1 \\ -2 \ 2 \end{matrix} , \begin{matrix} -1 \ 1 \\ -2 \ 2 \end{matrix} , \begin{matrix} -1 \ 1 \\ -2 \ 2 \end{matrix} , \begin{matrix} -1 \ 1 \\ -2 \ 2 \end{matrix} , \begin{matrix} -1 \ 1 \\ -2 \ 2 \end{matrix} \right)$$

Figure 2.13.9 Notating composite pleats.

Figure 2.13.10 Reorienting one pleat.

Figure 2.13.11 Stopping the pleat flattening early.

Figure 2.13.12 LEFT: Finished hex twist with flattened pleats front. MIDDLE: Finished hex twist with flattened pleats reverse. RIGHT: Pulling apart the twist.

a hexagonal backtwist and reorient one of the pleats.

Flatten the pleat from the edge toward the hexagon but stop two grid lines away from the hexagonal platform as shown in Figure 2.13.11.

What I really like about partial flattening is how it opens up the design to manipulation. You can see in the last photo of Figure 2.13.12 that it offers the folder the interesting option to pull the twist apart and introduces more depth to the piece. We'll explore this feature in Section 2.15 with the hidden circles pattern.

2.14 Triangle Twist Tessellation with Flattened Pleats

Now you'll take the concept of flattened pleats and apply it to an entire tessellation.

For this design, you'll start with a pattern of triangle twists (the 3^6 tessellation). The only difference is the triangle twists will be twice as wide as normal, allowing you to flatten the pleats on the grid more easily. Counting from triangle center to another triangle center,

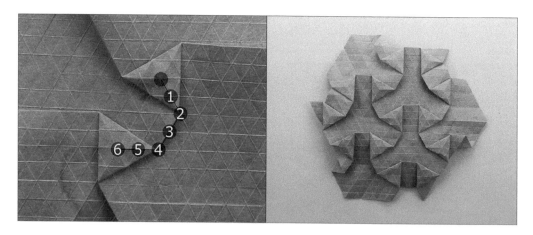

Figure 2.14.1 Starting tessellation, a 3^6 triangle twist tessellation with double-wide pleats and a spacing of six.

Figure 2.14.2 Incremental backtwisting.

make the spacing six grid lines and tile that throughout the paper.

Starting with a double-width triangle twist pattern that you will then manipulate and modify at a smaller subdivision introduces a multi-level tessellation design technique, which I call *scaffolding*. Scaffolding is used to describe a tessellation process where you fold a particular pattern first, and then use twist/pleat/floor modifications (such as flattening, backtwisting, reorientation, etc.) to create a different pattern without changing the overall structure. This leads you to understand a tessellation not just as a single diagrammable piece, but as a family of possible pieces, with a vast array of options left to the artist. As you learn more modifications, go back through the tessellations presented thus far, and see how you can change them to create more interesting effects.

For this piece, the modification you'll apply is the backtwist. However, you may notice that if you try to backtwist all of the triangle twists at once, the folds of the backtwists will clash, so you have to backtwist the center triangle first, retwist it to its original position, then backtwist each of the triangles surrounding it, retwist them, etc. Backtwist and retwist each triangle twist of the tessellation from the center outward.

When you have retwisted each of the triangles, there will be precreases not on the original grid that will help set up future steps (Figure 2.14.3).

Lift one of the pleats connecting two triangle twists as shown in Figure 2.14.4. Separate and push down in the center of the pleat, flattening it along the existing grid crease. While it is hidden in the second photo of Figure 2.14.4, it helps to place the index finger of your less dominant hand behind the paper inside of the pleat to separate it. Note that the twist will not lie flat yet!

Do this with each of the three pleats surrounding the starting triangle. Once you've finished all three, the pleats connecting the

Figure 2.14.3 Finished backtwisting.

Figure 2.14.4 Setting up the flattening of the pleats.

Figure 2.14.5 Finished flattening.

Figure 2.14.6 LEFT: Finished 3⁶ tessellation with flattened pleats front. RIGHT: Finished 3⁶ tessellation with flattened pleats reverse.

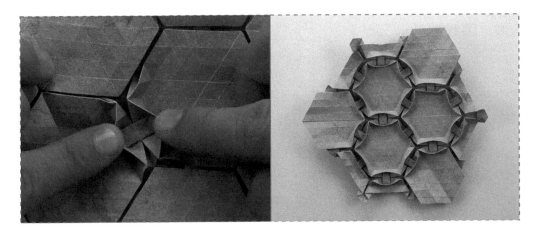

Figure 2.14.7 Decoration.

central triangle twist to its neighbors can be completely flattened and the center triangle will lie flat (though the triangles surrounding it won't). Keep working the pleats into flattened form outward. As you go, you can lay flat any twist that has fully flattened pleats.

When you've finished, you can leave the tessellation as it is in Figure 2.14.6; it's quite

a nice patterning effect. Alternatively, you can go onto the next step, which adds a decoration.

Peeling apart the edges of the hexagonal platforms on the reverse side reveals interesting bars underlying the pattern. I generally prefer to create the piece with some depth like the variant shown in Figure 2.14.7.

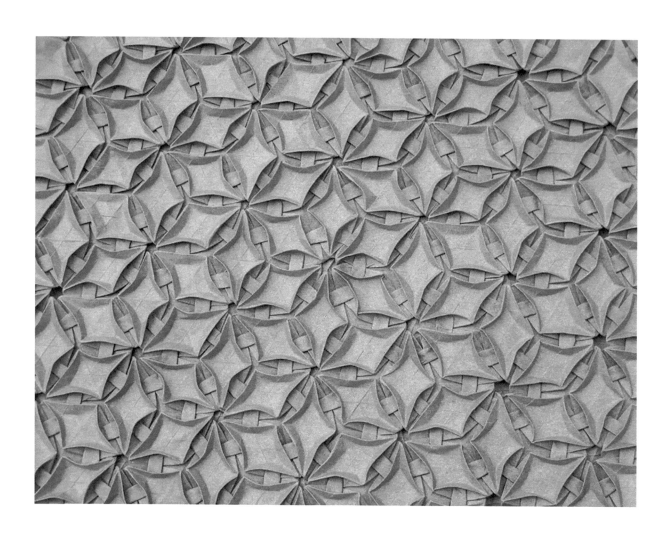

2.15 Hidden Circles Pattern

In the last section, I walked you through how to flatten a pleat and how to backtwist. The next tessellation, hidden circles, is one that really surprised me when I first folded it. Fairly straightforward in construction, it reinforces important concepts that become useful for other kinds of tessellation design, such as scaffolding, backtwisting, and pleat flattening. Its base pattern is a 3.6.3.6 twist tessellation with partially flattened pleats. As with the 3^6

pattern, you'll start with twists composed from double-wide pleats.

Figure 2.15.1 shows what the base tessellation looks like on a 32nds grid. In this case, there is a spacing of eight from the center of the hexagon to the center of the triangle twist. Note that this would only be a spacing of four if folded on a 16ths grid. Backtwist each of the molecules. Unlike the 3^6 pattern with the spacing shown in the last section, there is enough space between the molecules for them to all be backtwisted at the same time.

Figure 2.15.1 First tier tessellation and backtwist.

Flatten the pleats connecting the twists. On the side that approaches the hex backtwist, only partially flatten the pleats. The right photo of Figure 2.15.2 shows the finished tessellation on the front.

Figure 2.15.3 shows the reverse side. Open up the edges of the rhombic floors and see if you can figure out why the pattern is named *hidden circles*.

Figure 2.15.2 Partially flattening the pleats.

Figure 2.15.3 Decorating the reverse.

2.16 Rhombic Twist Tessellation

You've seen applications for the triangle and hex twists, as well as the triangle spread and hex spread twists. What are the uses for the rhombic twist? It follows the same ideas, one molecule's pleats flowing into the surrounding ones. Unlike the other twists I've talked about, pleats can extend from or flow into the rhombic twist in two main ways, from the pleat that is most CCW in a cluster of successive pleats—the leading pleat—or the pleat that is most CW in a cluster of successive pleats—the trailing pleat. To start to understand working with the rhombic twist, let's learn a tessellation that uses only the rhombic twist, simply called the *rhombic twist tessellation*, also called the rhombus weave [2].

Start with a rhombic twist in the center of the paper (like the triangle spread, it won't be exactly in the center).

Create a pleat parallel to one of the trailing pleats as shown in Figure 2.16.2; a trailing pleat of a rhombic twist is the more CW pleat in a pair of axes that contributes to the twist. When oriented toward the first twist's trailing pleat, both the new and the first pleats' mountain folds should align as shown in the second photo of Figure 2.16.2. Reorient the new pleat.

The steps in Figure 2.16.2 create a lock. Unlock that intersection and use the memory of the pleats to create a new rhombic twist rotating CW. We count these rhombic twists as having a spacing of three, counting from the border of the platform in standing form of one

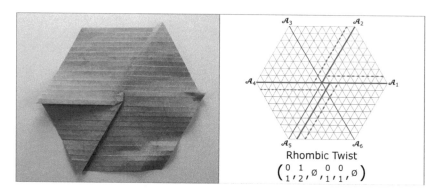

$$\text{Rhombic Twist}$$
$$\left(\begin{smallmatrix} 0 & 1 \\ 1 & 2 \end{smallmatrix}, \varnothing, \begin{smallmatrix} 0 & 0 \\ 1 & 1 \end{smallmatrix}, \varnothing\right)$$

Figure 2.16.1 Starting rhombic twist.

Figure 2.16.2 Setting up the second twist.

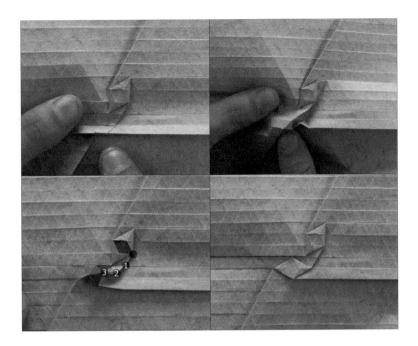

Figure 2.16.3 The second twist.

to the platform in standing form of the other, as shown in the third photo of Figure 2.16.3

Rotate the paper so the pleat connecting the two twists runs horizontally across the paper. Fold another pleat that will (as before) align with the pleat already present on the paper when the new pleat is oriented toward it. Reorient the new pleat (Figure 2.16.4).

This new pleat creates two locks. Resolve them one at a time. Figure 2.16.5 shows the resolution of one of the locks to create a CCW rhombic twist.

Continue this process of folding a parallel pleat, unlocking, and folding a new rhombic twist in the pattern shown. Figure 2.16.6 shows one possible progression of the pattern across the paper. Rotate after each set of unlocks. Each time you rotate the paper, the number of unlocks will increase by one.

Figure 2.16.4 Setting up the third and fourth twist.

Figure 2.16.5 Unlocking for the third twist.

Figure 2.16.6 Continuing the pattern.

Figure 2.16.7 LEFT: Finished rhombic twist tessellation front. RIGHT: Finished rhombic twist tessellation reverse.

When you have finished the process with the entire sheet, Figure 2.16.7 shows what the pattern looks like. Notice how one dimension of the paper is much more condensed than the other two. This is because the rhombic twist pulls the paper more from one direction than the other.

2.17 Rhombic Twist Variants

The rhombic twist gets more interesting when you reorient one or two pleats of the form. This leads you to new modifications of the rhombic twist, which I call the *kite* and *nub* molecules.

To create the kite molecule, start with a standing rhombic twist. Reorient one of the leading pleats in a cluster as shown. Then flatten the twist.

The resultant platform becomes a kite (Figure 2.17.1).

Figure 2.17.1 Reorienting a pleat of the rhombic twist.

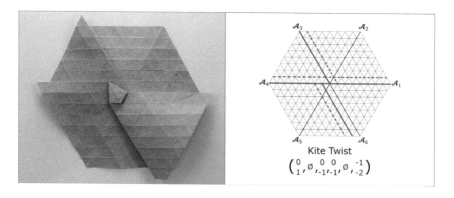

Kite Twist

$$\begin{pmatrix} 0 & & 0 & 0 & & -1 \\ 1 & \varnothing & {}_{-1} & {}_{-1} & \varnothing & {}_{-2} \end{pmatrix}$$

Figure 2.17.2 Finished kite twist and pleat schematic.

Figure 2.17.3 Reorienting the leading pleats.

The nub twist is similar but reorients both leading pleats. Press down from directly above to flatten the nub twist (Figure 2.17.3).

The reverse side gives a shifted appearance. You'll take advantage of that feature in the next tessellation. Also, notice how open the pattern can be peeled apart without unfolding the molecule. Like the hex twist with partially flattened pleats in Section 2.13, this molecule has some elasticity that can be interesting to play with.

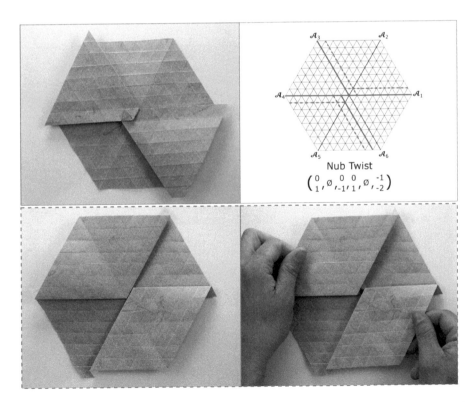

Figure 2.17.4 TOP LEFT: Finished nub twist front. TOP RIGHT: Pleat schematic. BOTTOM LEFT: Finished nub twist revers. BOTTOM RIGHT: Pulling apart the twist.

2.18 Twist Sinking

For the next tessellation, you'll encounter two more modifications. Just as backtwisting makes pleat flattening possible, it also makes *sinking* possible. A twist is *sunk* when its platform is on the opposite side of the paper from the pleats that form it. As with backtwisting, this is most easily done with twists that rotate about a point, such as the triangle, hex, and arrow twists, though it's most often used with just the triangle and hex twists.

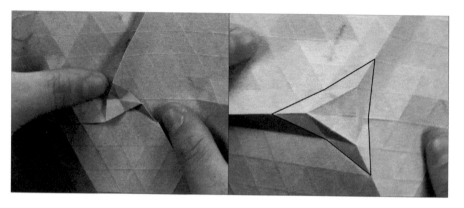

Figure 2.18.1 Unfolding the backtwist.

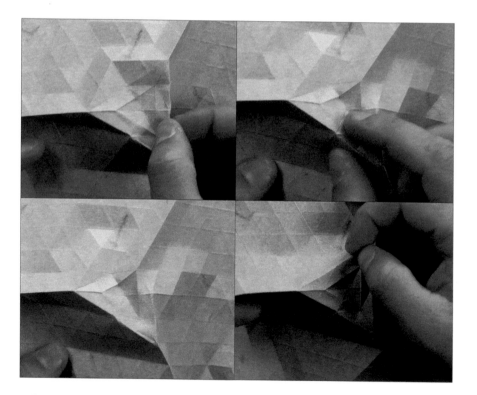

Figure 2.18.2 Pinching the border.

Figure 2.18.3 Closing up the sink. LOWER LEFT: Finished sunk triangle twist front. LOWER RIGHT: Finished sunk triangle twist reverse.

To sink a triangle backtwist, unfold it. Notice the border that is left as a result of the backtwisting (Figure 2.18.1).

A little bit at a time, pinch the border into a mountain fold until you've made a triangular platform with four grid triangle heights along each edge.

Figure 2.18.4 Unfolding the backtwist.

Dipping the center of the triangular platform away from you, pinch the border edges together towards the center and reform the original pleats around the triangle backtwist until the molecule lies flat again (Figure 2.18.3). This is a *sunk triangle twist*.

The hexagon analogue, the *sunk hexagonal twist*, follows a similar process. The difference is that the border created by the backtwist is already on the grid, so pinching the border of the expanded platform takes less negotiation than for a triangle twist sink.

Just as before, pinch the border of the hexagon twist (Figure 2.18.4).

Pinch the border together until the hexagon passes to the reverse side of the paper, then reform the original pleats and fold flat (Figures 2.18.5–2.18.6).

Figure 2.18.7 shows what the sunk hexagon looks like. You can orient the pleats so they rotate in the same direction for the full effect.

Figure 2.18.5 Forming the border.

Figure 2.18.6 Folding the molecule.

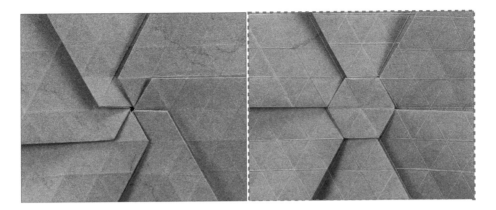

Figure 2.18.7 LEFT: Finished sunk hex twist front. RIGHT: Finished sunk hex twist reverse.

Figure 2.19.1 Sunk triangle twist.

2.19 Twist Expansion

The sink leads neatly into the next modification, twist expansion. The backtwist makes a secondary border which is hidden under layers of paper. The sink takes that border and pushes everything inside of it to the reverse side of the paper. Expansion makes that border the edge of the molecule itself.

For a triangle expansion, start with a sunk triangle twist.

Peel apart the border until the triangle lies flat on the paper. Doing so forces an extra crease to be created in the corners (shown with the arrow in the second photo of Figure 2.19.2).

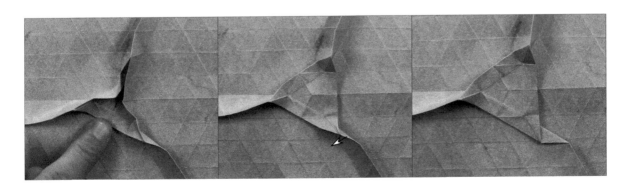

Figure 2.19.2 Opening the twist.

Figure 2.19.3 shows the finished expanded triangle twist. Notice how the pleats are the same as the triangle twist (and would have the same pleat intersection notation), but the molecule encompasses a larger area than a regular triangle twist.

The hex twist expansion is a little more straightforward. Just peel apart the borders of a sunk hex twist until the larger hexagon lies flat on the table. No new creases are required.

Figure 2.19.5 shows the finished hex twist expansion.

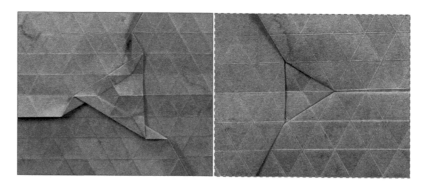

Figure 2.19.3 Finished expanded triangle twist. LEFT: Front. RIGHT: Reverse.

Figure 2.19.4 Opening a sunk hex twist.

Figure 2.19.5 Finished expanded hex twist. LEFT: Front. RIGHT: Reverse.

2.20 Nub Offset Tessellation

The nub twist mentioned in Section 2.17 is rather odd compared to the pleat intersections I've looked at so far. Whereas the six simple twists have their pleats all lying flat in either a CW or CCW direction, the nub twist has two pleats oriented in one rotation and two in the other. You'll now make use of the nub twist—as well as twist sinking and expansion from Sections 2.18 and 2.19—in a full tessellation.

To start the nub offset tessellation, you'll begin with a sunk hex twist in the middle of the paper. The benefit of this central molecule is that it has extra paper that can be freed, allowing you to turn the corners of the hex twist into nub twists, one at a time.

Reorient one of the pleats to get it out of the way. You're going to split the pleat that it moved away from, but in a way that will turn it directly into a nub twist. Pinch as shown in the final photo of Figure 2.20.2.

Once the rhombic platform has been created, lay the pleats flat to the edge of the grid.

Do the same with the second pleat. The nub twist's pleats will be oriented in the same

Figure 2.20.1 Sunk hex twist.

Figure 2.20.2 First molecule setup.

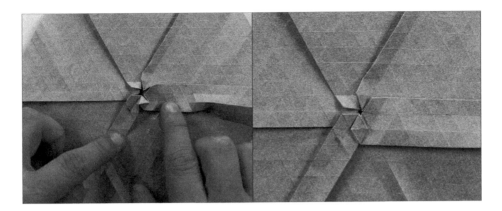

Figure 2.20.3 First nub molecule.

direction throughout this tessellation. The second nub twist around the starting hexagon (and each nub twist hereafter) will create a lock. Keep folding nub twists until you have a cluster of six nubs surrounding the starting sunk hex (Figure 2.20.4).

Figure 2.20.4 First cluster.

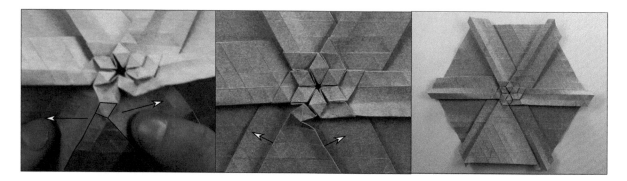

Figure 2.20.5 Second row of rhombic twists.

Now, you'll deal with one of the locks. Unlock the pleats as before, and find the rhombic platform from the split to create the next nub twist (second photo of Figure 2.20.5). This one should fit snugly with the two that created the lock.

Keep working your way around the first cluster, unlocking pleats. Figure 2.20.6 shows what it looks like when the first set of locks have been resolved.

To get the next cluster, notice that there are two simple pleats on each axis, each oriented toward each other (Figure 2.20.7). Hold those pleats a few triangles away from the first cluster and open them up. The acute corners of the three nub twist platforms all point to single grid intersection. Pinch the border of the hexagon surrounding that point and create a sunk hexagon.

This new sunk hex molecule gives us the ability to continue to pattern from there. Now you can continue the unlocking and twisting. Notice areas where there are pleats to unlock. Sometimes you will have to unlock multiple at once, as in the first photo of Figure 2.20.8.

Tile like this until you run out of paper. Figure 2.20.9 shows the finished tessellation.

This particular pattern has a decoration I rather enjoy. To create this decoration, press

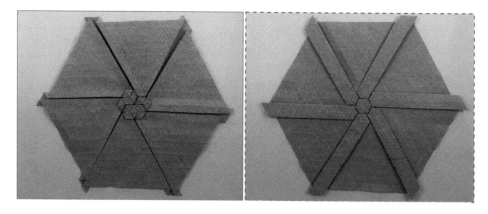

Figure 2.20.6 First and second rows.

Figure 2.20.7 Second cluster setup.

Figure 2.20.8 Second cluster finished.

Figure 2.20.9 Finished nub offset tessellation. LEFT: Front. RIGHT: Reverse.

Figure 2.20.10 Inversion decoration.

on one of the hexagons on the reverse side, and open the molecule (Figure 2.20.10).

Identify the six grid triangles in the center, which will form a hexagon, and press down on the outside of this hexagon. It can be helpful to use a tool like the one shown in Figure 2.20.11 (a stylus). When you have pressed the paper surrounding the hexagon away, you will have a hex twist on this side of the paper instead and an interesting contrast to the rest of the piece. You can invert any number of hexagons, leading to a lot of choice in the patterning.

Figure 2.20.11 Inversion decoration continued.

Figure 2.20.12 Finished inversion decoration. LEFT: Front. MIDDLE: Reverse. RIGHT: More decoration.

2.21 Shift Rosette Tessellation

This shift rosette tessellation makes use of the hex twist surrounded by nub molecules. The combination creates a neat honeycomb pattern.

For this pattern, you'll start with a hex twist in the center (Figure 2.21.1). Split one of the pleats four spaces away from the hex center (the mountain folds of the two parallel pleats will line up as shown).

When you make the second split, there will be a lock (first photo of Figure 2.21.2). Go ahead and resolve the lock right now rather than waiting like you've done in other tessellations. Unlock the pleats and mold the outline of a hexagon, similar to the outline of a sunk hex (but only four sides). You are essentially resolving multiple splits simultaneously. The result will lie flat as shown.

This creates one petal of the honeycomb on the reverse side of the paper, as you can see in the first photo of Figure 2.21.3. Flip back to the front and continue splitting and molding the outline until the full cluster of splits and moldings are finished.

Now you'll turn one of the corners of the molding into a nub twist. Start by opening the pleat. Find the rhombus adjacent to the corner and push the pleats in until the rhombus platform twists around (final photo of Figure 2.21.4).

That first nub creates a new lock with the first pleat of the next split. Unlock those pleats and twist the next nub in the same way as before

Figure 2.21.1 Starting hexagon and a split four spaces away.

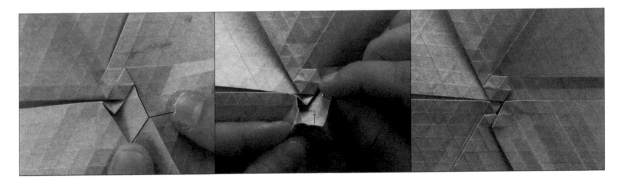

Figure 2.21.2 Second split and ridgeline.

Figure 2.21.3 Continuing the splits.

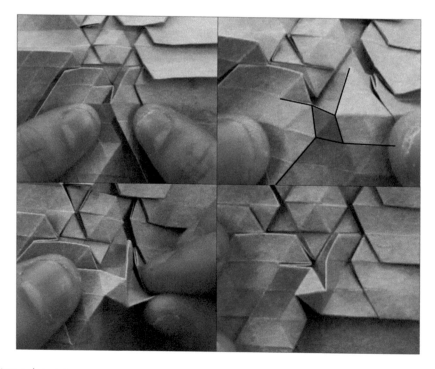

Figure 2.21.4 First nub twist.

Figure 2.21.5 S-platform creation.

(Figure 2.21.5). Two pleats will collide, and you can combine them to create an S-shape—highlighted in the first photo of Figure 2.21.6.

Flattening the nubs reveals the S-shape in its entirety. Flip the paper to reveal an interesting shifting effect that comes from the S-shape on the reverse side.

Fold the second S-shape coming off of the center hexagon. When you have finished, notice that there are two nub molecules and a lock (first photo of Figure 2.21.7). That lock will create a third nub twist in a cluster.

Unlock the pleat as shown in Figure 2.21.8 and find the nub rhombus platform that will fit the third nub molecule of the cluster and flatten.

Keep folding these S-shapes around the center hexagon and the nub molecule clusters until the center molecule is surrounded by them (Figure 2.21.9).

The center hexagon, the S-shape, and the nub molecules combine to create a larger, more complicated molecule. Next, you'll have to find where the next molecule will start. For this, you split one of the pleats coming of the corner of an S-shape as marked in the first photo of Figure 2.21.10 and count three grid lines away to find where the next hex twist will be centered.

Split the pleat coming from that S-shape. This point will be a corner of the hexagonal platform of the hex twist of the second molecule. This creates two locks; resolve them one at a time

Figure 2.21.6 S-platform finished. LEFT: Front. RIGHT: Reverse.

Figure 2.21.7 Nub cluster, second twist.

Figure 2.21.8 Finishing the nub cluster.

Figure 2.21.9 First iteration finished. LEFT: Front. RIGHT: Reverse.

Figure 2.21.10 Finding the next hexagon.

into splits, with pleats that converge on where the hex twist will end up (fourth photo of Figure 2.21.10).

Where they converge, turn that into the center of a hex twist (Figure 2.21.11). This hex twist and the initial hex twist have one S-shape between them and two nub molecule clusters, the S-shapes being marked in the third photo in Figure 2.21.11.

Do the same with each S-shape and tile the paper with the next cycle of hex twists. Figure 2.21.2 shows the finished tessellation.

Figure 2.21.11 Finding the next hexagon continued.

Figure 2.21.12 Finished shift rosette tessellation. LEFT: Front. RIGHT: Reverse.

2.22 Ridge Creation

Ridge creation is another useful skill for more complex tessellation folding. A *ridgeline* or just *ridge* is an elevated mountain fold path that follows grid lines and is surrounded on either side by valley folds. In other words, it is a standing pleat, and connecting it to other standing pleats will allow you to create a network of these ridges. But there is a challenge here; whereas a flat pleat can be folded, and you can have confidence it will stay folded, a ridge relies on manipulating the paper memory to keep the form in place. The harder and

Figure 2.22.1 Starting ridgeline formfcation.

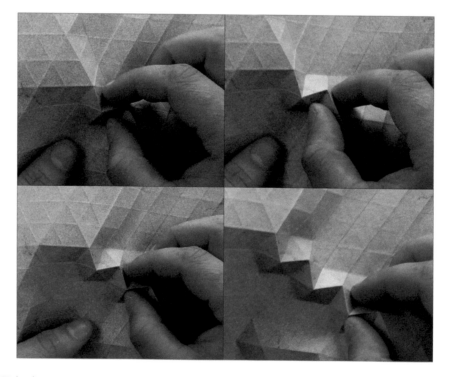

Figure 2.22.2 Ridgeline continuation.

Figure 2.22.3 Ridgeline cycle.

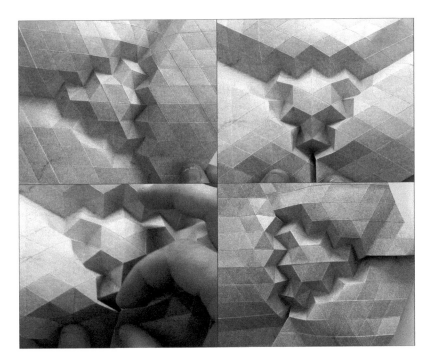

Figure 2.22.4 Turning a ridgeline cycle into a pleat intersection.

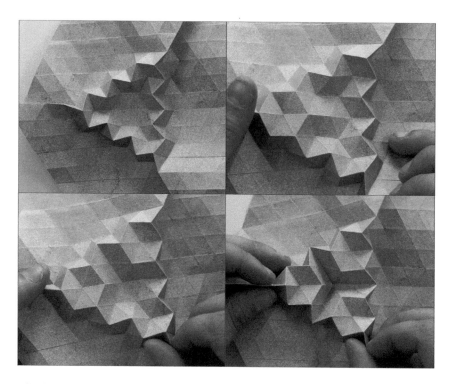

Figure 2.22.5 Finished ridgeline—built molecule.

longer you pinch the mountain fold of a ridge, the more memory that portion of the ridge will have, and the more detail you can create.

To form a ridgeline, pinch on either side of a path following grid lines. This will force the mountain of the standing pleat off the table. You can sculpt from there.

The photos in Figure 2.22.2 show how to make a ridgeline that creates a stair step pattern.

Once you've created an entire shape from the ridgeline, such as the triangular form in the third photo of Figure 2.22.3, you can extend one of the ridges to the edge of the paper.

Once these extensions get far enough away, you can treat them as normal flat pleats that can flow into a new molecule or you can make a new ridgeline from there (Figure 2.22.4).

The interesting thing is that you really have the ability to sculpt as you wish with ridges, leading to a vast number of possibilities. The set of photos in Figure 2.22.5 shows a transition between ridges and simple pleats at the corners of the ridge with some decoration.

2.23 Button Molecule

The button molecule is a modification of the hex spread that creates a tessellation that is quite open and interesting. It makes use of the memory of the paper to keep the form together, like the ridges described in Section 2.22. Start with a hex spread twist in standing position.

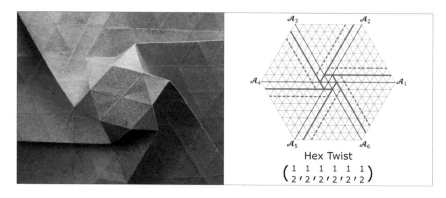

Figure 2.23.1 Hex spread twist in standing form.

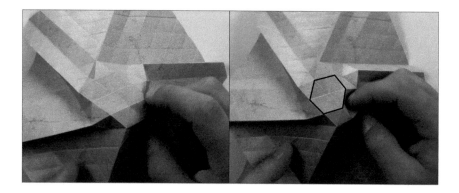

Figure 2.23.2 Popping in the corners of the hexagon.

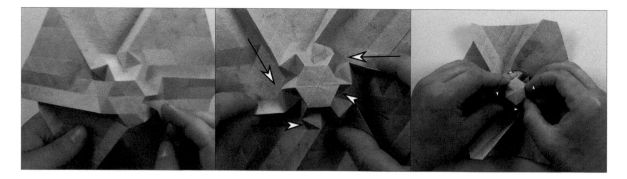

Figure 2.23.3 Twisting the molecule.

138

Figure 2.23.4 Finished button molecule. LEFT: Front. RIGHT: Reverse.

Put your fingers behind the molecule and partially unfold it. Notice the hexagon in the middle of the molecule—highlighted in the second photo of Figure 2.23.2. Push in on the corners of the molecule one at a time until the edges of the top platform are the six-triangle hexagon instead of the larger one.

Finish pushing each of the pleats in the same manner. When you have done all six, use the pleats to push the hexagonal platform off the table. You can then twist the hexagon the opposite rotation from the pleats and lay the form "flat (Figure 2.23.3)."

Figure 2.23.4 shows the finished button molecule. It doesn't lie flat, but the memory of the folds keeps the form together. In the next section, you'll tile these on the paper with ridge molecules between.

2.24 Button Molecule Tessellation

To tile the button molecule, start by sculpting the ridgeline in a stair step pattern as shown in Figure 2.24.1.

Sculpting further, you can get the triangular ridgeline that you learned in Figure 2.23.2. The third photo in Figure 2.24.2 shows where the hexagon platforms of future hex twists will be. Form more triangular ridgelines offset from one another by one space as shown.

Now that you've identified where the second row of hex twists will be, you can get the pleats that create that twist in place. Turn the ridgelines on the sides into flattened pleats, keeping in mind that they aren't actually going

Figure 2.24.1 Extending the ridgeline.

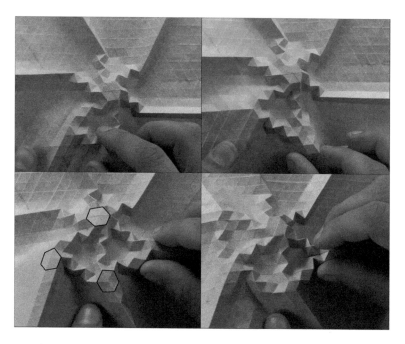

Figure 2.24.2 Setting up the second row of button molecules.

to be flat due to their connection to standing pleats. The end of each of these flattened pleats closer to the first molecule will be where another hexagon goes (third photo of Figure 2.24.3).

Turn one of those flattened pleats into a button molecule (Figure 2.24.4).

Keep tiling the buttons in this way. In the photos of Figure 2.24.5, I do a cluster of four buttons. After you have a cluster, you can grab the hexagon platforms of the buttons and pinch them together to increase the memory of the state the paper should be in. Continue the pattern to the edges, pinching occasionally as necessary.

Figure 2.24.6 shows the finished button molecule tessellation. It's not a flat tessellation like the others you have done, but it is helpful to see other forms tessellations can take to develop stronger design skills.

Figure 2.24.3 Setting up the second row of button molecules continued.

Figure 2.24.4 Finishing the second molecule.

Figure 2.24.5 Finishing a cluster of four molecules.

Figure 2.24.6 Finished button molecule tessellation. LEFT: Front. RIGHT: Reverse.

2.25 Triangle Flagstone Tessellation and Offsetting Pleats

In the origami tessellation community, flagstone tiling has long been a source of interest. The draw of the flagstone comes from the "reverse" side of the paper, the side that shows the floors, rather than the pleats. Due to offset patterning, the backside floors are rotated with regard to the grid, which looks like a flagstone tiling on a sidewalk.

Standard molecule tiling involves creating a twist and using the pleats that come from it to create new twists, and from that, a pattern is generated. But what happens if two twists are slightly offset from one another, where the mountain folds of the pleats, while parallel, do not line up with each other? This is an offset tiling, and creates that motif described previously.

On the right side of Figure 2.25.1 the mountain folds of the twists do not line up. The paper will allow for a certain amount of correction, and the mountains divert to connect the corners of the triangle platforms. The dotted blue line shows this diversion.

To understand this new skill of offsetting, begin with a CCW triangle twist in the paper. Unfold the twist and drift one of the pleats CW one grid line. There will be a puffed rhombus— shown in the middle image of Figure 2.25.2

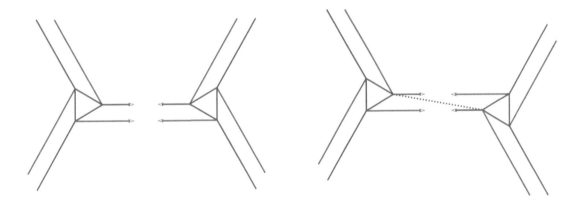

Figure 2.25.1 LEFT: Normal molecule connection. RIGHT: Offset molecule connection.

Figure 2.25.2 Drifting a pleat from a triangle backtwist.

143

Figure 2.25.3 Splitting the offset pleat.

Figure 2.25.4 Turning the split into a new triangle backtwist.

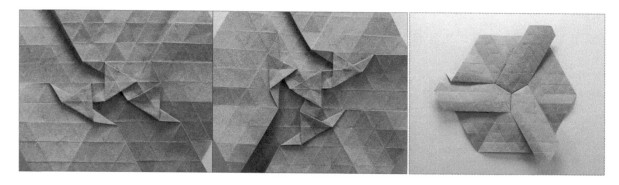

Figure 2.25.5 Continuing the pattern. RIGHT: Reverse view after the second row.

Figure 2.25.6 Finished 3^6 flagstone tessellation. LEFT: Front. RIGHT: Reverse.

that you can compress. The fully drifted pleat is shown in the right photo.

Then split the drifted pleat one space away from the drift (middle image of Figure 2.25.3).

Once you've split the pleat, you can create a triangle twist that is similar to a triangle backtwist in that it twists in the opposite rotation for two of the pleats (Figure 2.25.4). Press down on the corners of the triangle to flatten it. Do not worry if the triangle looks inaccurate the first few times you do this. This is not a natural setup for the triangular twist and takes a lot of practice!

Now you have finished the offset. Split the other pleats of the initial triangle twist and create the next iteration of offset triangle twists (Figure 2.25.5).

Continue this process of locking, unlocking, offsetting, splitting and flattening until you have tiled the entire grid. Figure 2.25.6 shows the finished tessellation.

2.26 3.6.3.6 Flagstone Tessellation

Another method for folding flagstones uses the button molecule, almost creating ridgelines and using them as guides to create the offsets. We'll use this method for the 3.6.3.6 flagstone tessellation. For this, start with a button molecule in the center of the paper.

Pinch the ridgeline where the pleat meets the molecule as shown in Figure 2.26.2. Do this to all six pleats and lay the pleats down in a CCW rotation. The hexagon platform will twist flat.

Split one of the pleats and start to rotate it CCW, the same rotation as the hex twist. You'll notice that it won't lie flat just yet (Figure 2.26.3).

Figure 2.26.1 Button molecule formed by offsetting pleats from a hex twist.

Figure 2.26.2 Flattening the hex twist with offset pleats.

Rotate the triangular platform CW instead, and the twist will lie flat. Now backtwist the triangle platform.

Repeat the triangle twist and backtwisting around the hexagon, and there will be six locks that will be resolved soon (Figure 2.26.5). Flip the paper, and you'll find the start of a rhombille tiling, the dual of the 3.6.3.6 tiling.

Unlock the pleats, identifying a bar with a width of one at the intersection (Figure 2.26.6).

Figure 2.26.3 Splitting one of the offsets.

Figure 2.26.4 Turning the split into a triangle twist.

Figure 2.26.5 First row of triangles. RIGHT: Reverse.

Figure 2.26.6 Finding the next hexagon.

Before flattening, offset the formerly locking pleats CCW so that they line up with the offsetting of the triangle twists that form them (Figure 2.26.7).

This shows the offsetting of the unlock. Flatten that unlocked bar and repeat with all six locks. The unlocked bar forms the basis for a new button molecule (Figure 2.26.7). Start to form another offset pleat to flow into the new button molecule.

Finish the second button molecule and use the steps in Figure 2.26.8 to flatten into a new hex twist. There will be several locks that can be resolved later.

Figure 2.26.7 Turning the split into a bar with a width of one. BOTTOM-RIGHT: Starting the next button molecule.

Figure 2.26.8 Forming the second hexagon.

Figure 2.26.9 Setting up the third hexagon.

Figure 2.26.10 Finished second row of hexagons. RIGHT: Reverse.

One of those locks unlocks into three pleats of a third hex twist, between which we can create a new triangle twist. Follow the steps in Figure 2.26.9 to continue the tiling, unlocking as needed.

Figure 2.26.10 shows the tessellation when the first ring of hex twists is finished and the rhombille tiling on the reverse is more apparent.

Figure 2.26.11 shows the finished 3.6.3.6 tiling.

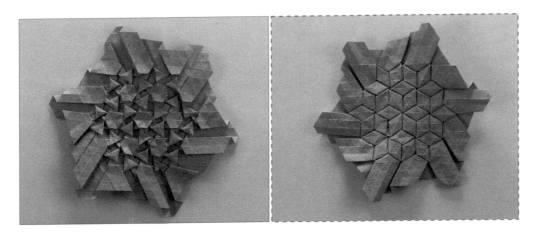

Figure 2.26.11 Finished 3.6.3.6 flagstone tessellation. LEFT: Front. RIGHT: Reverse.

2.27 Crooked Split

The crooked split is an easy way to transfer from a flat tessellation to one with much more depth. To start, create a pleat split, but do not lay it flat.

Along one of the angles formed by the mountain folds, push the point of the split inward as shown in Figure 2.27.2.

The potency of this split is seen when connected with a hex twist. Take one of the

Figure 2.27.1 Simple pleat split.

Figure 2.27.3 Crooked split reverse.

Figure 2.27.2 Crooked split ridgeline.

Figure 2.27.4 Crooked split off a hex twist.

Figure 2.27.5 Finished crooked splits.

pleats of the hex twist and create a crooked split. Do this one at a time around the hexagon. Pleats will lock as they reach the hexagon edges, but on a 16ths grid, there isn't enough room for them to actually collide (Figure 2.25.4).

Figure 2.25.5 shows what the crooked splits look like when each of the splits are finished.

You can unlock those pleats and reform the location where they intersect as shown in Figure 2.25.6. Sculpt the ridgelines and lay the pleats in bars with widths of two.

This is the finished shaping step. You can leave it like this or flatten, with different esthetic effects for either choice.

Figure 2.27.6 Unlocking the pleats.

Figure 2.27.7 Flattening the snowflake.

Figure 2.27.8 Flattening the snowflake.

From the reverse side, you can flatten the "bowtie" shapes in the middle of the sides and the triangle forms in the corners, as shown in Figure 2.27.8.

Figure 2.27.9 shows the entire reverse side flattened.

If you really want the form flattened, there are two more items that we can flatten: the bars (Figure 2.27.9) and the splits (Figure 2.27.10).

Figure 2.25.12 shows the snowflake molecule, fully-flattened. You will use this in a full tessellation in Section 2.28.

Figure 2.27.9 Finished flattening reverse.

Figure 2.27.10 Flattening the bars.

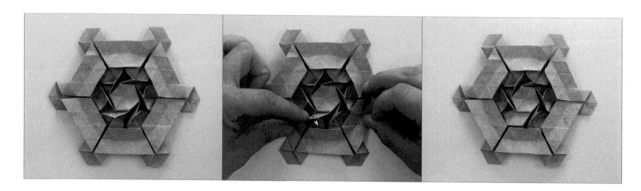

Figure 2.27.11 Flattening the crooked splits.

Figure 2.27.12 Finished snowflake molecule. LEFT: Front. RIGHT: Reverse.

2.28 Snowflake Tessellation

To make the snowflake tessellation, you'll start with the snowflake molecule learned in the last section. You can start with it in a flattened or three-dimensional state, though recognize that these photos will show the flat version.

To find the next molecule, open up one of the pleats as shown in Figure 2.28.1. Create the crooked split from the opposite side. This will create a new simple pleat going toward the corner of the paper.

To create the second hex, split the simple pleat two gridlines away from the center of the crooked split. At that split create the next hexagon with a CCW rotation (with all of the locks that go with it). Finish the crooked splits around the second hexagon (second photo of Figure 2.28.3).

Where the pleats from the first and second twist intersect, unlock all the pleats at the same time (third photo of Figure 2.28.3). Where the mountain folds intersect, form two crooked splits that point away from each other (fourth photo of (Figure 2.28.3).

Use the crooked split formed in Figure 2.28.3 as a guide to find the other crooked splits that will form the third molecule (Figure 2.28.4). Where the pleats from the crooked splits intersect, form the third hexagon. This will create another set of several locked pleats. Unlock and then create the crooked splits as before.

Continue this pattern until the entire piece is tiled. Figure 2.28.5 shows the finished tessellation.

Figure 2.28.1 Forming the split of the second molecule.

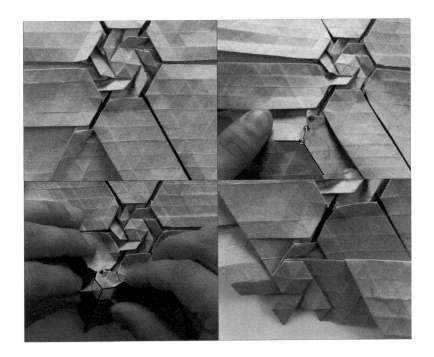

Figure 2.28.2 Finding the second twist.

Figure 2.28.3 Finding the third molecule.

Figure 2.28.4 Finished third molecule.

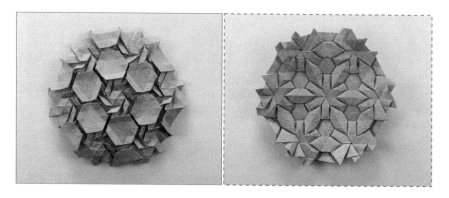

Figure 2.28.5 Finished snowflake tessellation. LEFT: Front. RIGHT: Reverse.

2.29 Tulip Split

This is another split method that I call the *tulip split*. This is a method of transitioning between flattened and simple pleats in the same design.

To fold the tulip split, start with a flattened pleat.

Pull apart one end of the flattened pleat. Identify the two marked triangles in the second photo of Figure 2.29.2. Those will rotate inward, toward the middle grid line of the still-flattened side of the pleat. The unflattened part will split into three separate single-wide pleats as shown. Those triangle platforms will connect on the corners.

Figure 2.29.1 Flattened pleat.

Figure 2.29.2 Tulip split setup.

Figure 2.29.3 Flattening the tulip split.

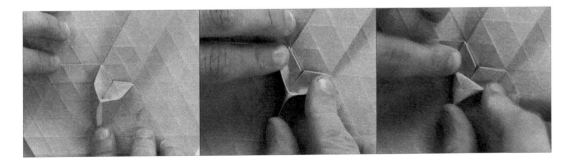

Figure 2.29.4 Flattening the triangle twist.

Now, lay one of the platforms flat. You can use your fingers, but sometimes I find it easier to use a pointed tool such as a stylus to help the paper into place (Figure 2.29.3).

The remaining part of the split can turn into a triangle twist. Lay that flat as shown in Figure 2.29.4.

Figure 2.29.5 shows the finished tulip split.

You can connect multiple tulip splits together to get interesting effects on the reverse (Figure 2.29.6).

For practice, I recommend trying to fold the tulip split from pleats connected to a triangle twist with flattened pleats.

Figure 2.29.5 Finished tulip split. LEFT: Front. RIGHT: Reverse.

Figure 2.29.6 Finished double tulip split. LEFT: Front. RIGHT: Reverse.

Figure 2.29.7 Tulip split from a triangle twist with flattened pleats.

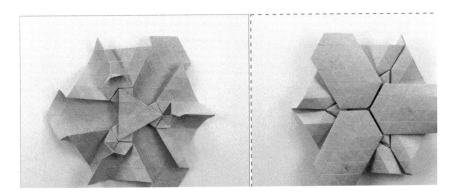

Figure 2.29.8 Tulip split from a triangle twist with flattened pleats. LEFT: Front. RIGHT: Reverse.

Figure 2.29.8 shows the result of creating the tulip splits from such a molecule; the offset decoration on the reverse is more apparent in a cluster.

2.30 Tulip Split Tessellation

I rather enjoy the tulip split coming off the hexagon, and this can be used to create a tessellation. You'll start with a hex backtwist with double-wide pleats. Reorient one of the pleats and begin to flatten the pleat adjacent to it, but don't flatten fully (Figure 2.30.1).

Turn that flattened pleat into a tulip split. Flatten the intersections of the split as you did before and repeat around the hexagon. A lock will appear where each of these splits intersect (Figure 2.30.2).

Continue the flattening and resolve each of the locks as bars with widths of one. This creates a quite beautiful molecule with spikes on the reverse (Figure 2.30.3).

Unfortunately, you will have some difficulty tiling this again on the paper. There's just not enough room with thirty-two grid lines. How do you know?

Counting from the center of a hex twist to the center of a triangle split attached to a

Figure 2.30.1 Double-wide hexagonal backtwist.

Figure 2.30.2 Making the tulip splits.

Figure 2.30.3 Finished tulip split molecule. LEFT: Front. RIGHT: Reverse.

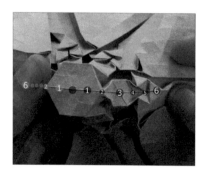

Figure 2.30.4 Determining the size of the molecule.

tulip split we see that it is six grid lines away. The counting is helped by partially unfolding the molecule as shown in Figure 2.30.4. The molecule extends in both directions from the center, as the brown dots represent. This results in a total molecule diagonal of twelve grid lines, which simply does not fit on the paper—fitting three of these across the diagonal would require thirty-six grid lines. The next section will give us some options to account for lack of space.

164

2.31 Molecule Size and Different Grid Densities

Figure 2.31.1 shows why a molecule with a diagonal of twelve grid lines do not fit on the 32nds grid.

If you liberate yourself from having to start in the center, you can fit four such molecules by starting from a point not on the diagonal (Figure 2.31.2).

If you had more grid lines, you'd be able to fit more molecules into your pattern. You can get more triangles if you extend your grid subdivisions to 64ths (Figure 2.31.4).

I encourage you to practice folding a 64ths grid using the same method as the 32nds from Section 2.1. Then you can try one or more of the tessellations on a denser grid.

You may notice that if you had just four extra grid spaces across each diagonal, you could fit another ring of molecules, so perhaps you should go back and refold it from a 36ths grid. The second photo of Figure 2.31.6 is a possible result. However, thirty-six isn't an easy division to find. One method is to get as close to 9ths as possible, and then subdivide each of the nine divisions into quarters. But how do you subdivide a grid into 9ths? Tom Full presents a solution in *Project Origami*, where he references a technique developed by Shuzo Fujimoto, the *Fujimoto Approximation* [3].

To start, I want to introduce an alternate method of folding the hexagon. This alternate method requires more practice but does not have the ¼ mark when you are finished. Begin it the same way as you usually do. Fold the rectangle in half, and then turn it and make a pinch mark at the midpoint not on the raw edge of the paper (Figure 2.31.7).

Fold approximately 60° without guides. After practice, you will get a feel for how this should be, but even after a lot of practice do not crease down yet. Just press enough to have a mark. Then flip the paper and fold the other flap to match. Again, you won't have guides so press enough to make a mark and no more. Work both creases flat at the same time so that there are three 60° angles located at the center mark of the

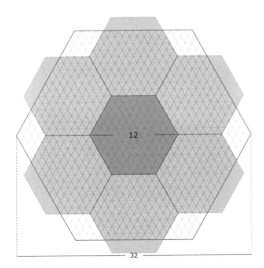

Figure 2.31.1 Fitting a molecule with a width of 12 spaces on a 32nds grid.

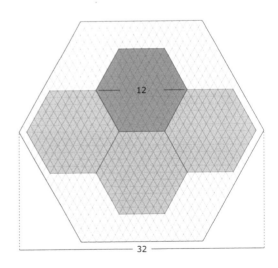

Figure 2.31.2 Alternative arrangement of the molecule on the grid.

Figure 2.31.3 Tulip split molecule with alternate assortment.

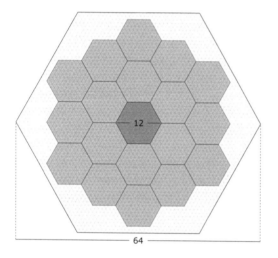

Figure 2.31.4 Fitting a molecule with a width of 12 spaces on a 64ths grid.

paper (Figure 2.31.8). Finish making the hexagon as normal from there as you usually would.

Once you have the hexagon, you will need more guides to assist in subdividing into 9ths. These will come at the bisectors of each of the diagonals. Fold a corner of the hexagon to the opposite corner and unfold. Repeat with all three pairs of opposite diagonals, and then backcrease each of those folds (Figure 2.31.9).

Now you need a visual of ⅛ of the paper for reference. Fold the edge closest to you to the halfway point and pinch the middle of the crease, lining up the indicated lines as you do so. Do not fully crease it. This allows you to visualize the ¼ mark. Then unfold and fold the same edge to the ¼ mark, pinching again and lining up the indicated lines. This allows

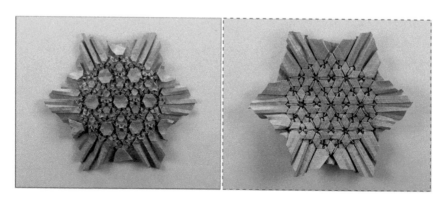

Figure 2.31.5 Tulip split molecule tiled on a 64ths grid.

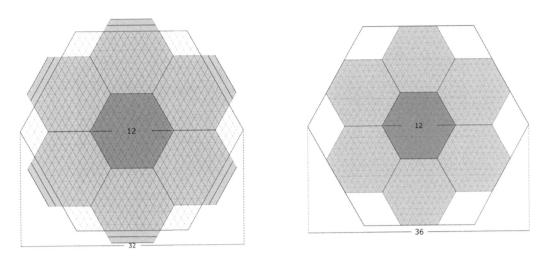

Figure 2.31.6 Fitting a molecule with a width of 12 spaces on a 32nds and 36ths grids.

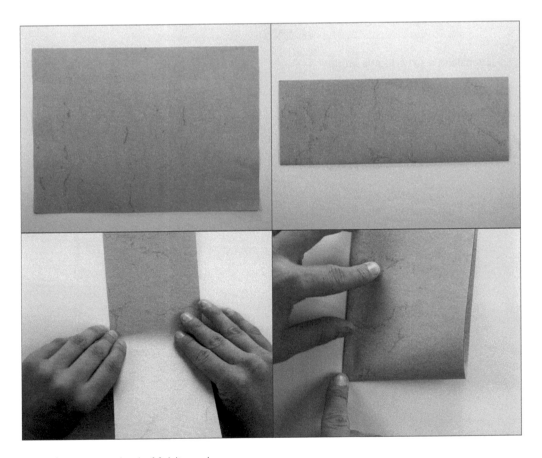

Figure 2.31.7 Alternate method of folding a hexagon.

Figure 2.31.8 Eyeballing a sixty degree angle.

Figure 2.31.9 Finding the perpendicular axes.

Figure 2.31.10 Using the perpendicular axis as a guide to visualize 8ths. BOTTOM-RIGHT: Approximating 9ths.

you to visualize the ⅛ mark. Then unfold and refold the edge to a little less than the ¼ mark, again lining up the indicated marks. The exact distance doesn't matter much but take your best guess at where a 1/9 crease would be (Figure 2.31.10).

Now, with this approximation technique, the second mark made is roughly 1/9 of the way up, which means the remainder of the paper is 8/9 of the paper (Figure 2.31.11). That 8/9 is easily subdivided into eight parts. Rotate the paper so the approximation marks are away from you.

You still want to just pinch because you're not sure if the creases you make are accurate divisions. Fold from the closest edge to you to the assumed 1/9 mark to find the

Figure 2.31.11 Guessing 1/9 on the grid.

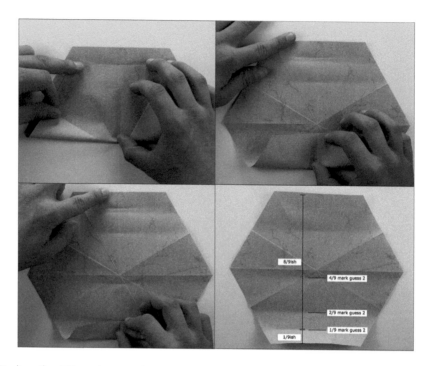

Figure 2.31.12 Finding the 1/9 mark second guess.

assumed 4/9 mark, pinching to mark the spot (Figure 2.31.12). Then pinch from the same edge to that pinch, finding the assumed 2/9 mark. Then pinch from the same edge to that pinch, finding the assumed 1/9 mark, second guess.

Repeat this process one last time, and the final 1/9 guess will be surprisingly close to the true 1/9 mark. The more iterations of this technique you perform, back and forth, the closer your 1/9 guesses will get to the true 1/9 mark. After three times, your creases should be certainly accurate enough to start making full creases (Figure 2.31.13).

We can subdivide the remaining 8/9 from there.

Figure 2.31.13 Finding the 1/9 mark final guess.

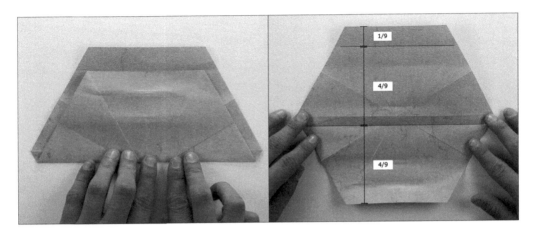

Figure 2.31.14 Subdividing the 9ths.

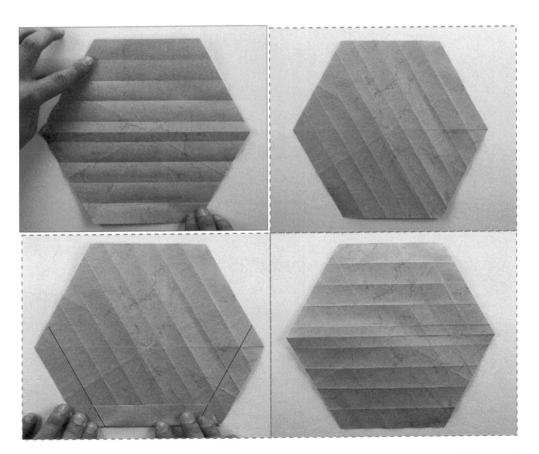

Figure 2.31.15 TOP LEFT: Finished first axis 9ths. TOP RIGHT: First axis 9ths backcreased. BOTTOM LEFT: Finding the second axis 1/9 mark. BOTTOM-RIGHT: Subdividing the second axis to 9ths.

Figure 2.31.16 Finding the third axis 1/9 mark. RIGHT: Subdividing the third axis to 9ths.

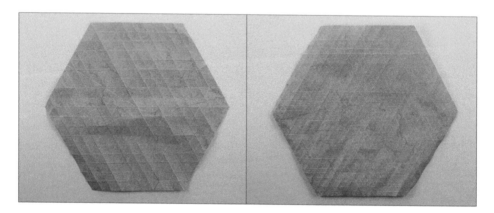

Figure 2.31.17 LEFT: 18ths grid. RIGHT: 36ths grid.

Figure 2.31.18 Tulip split molecule tiled on a 36ths grid.

Figure 2.31.19 3.6.3.6 tessellation using the tulip split molecule.

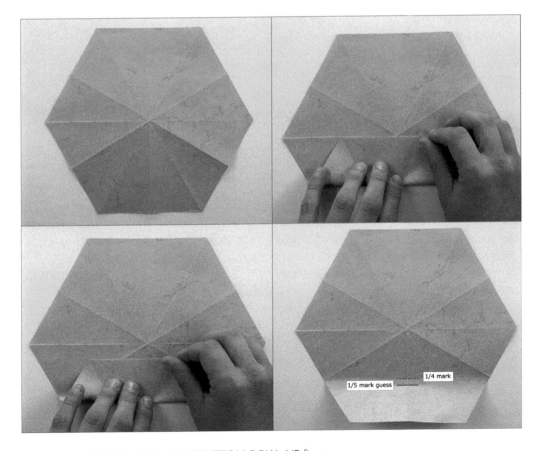

Figure 2.31.20 TOP ROW: Visualizing 1/4. BOTTOM ROW: 1/5 first guess.

173

Figure 2.31.21 Subdividing axes to fifths.

Figure 2.31.22 LEFT: 10ths. MIDDLE: 20ths. RIGHT: 40ths.

As usual, you should backcrease each of the 1/9 marks. Then rotate and use the first axis as a guide to find the starting creases for the second axis. The 1/9 mark on the second axis can be found by lining up the black lines in the third photo of Figure 2.31.15.

Likewise, you can use the first and second axis to find the 1/9 mark on the third.

Finishing each level of subdivision will give you 18ths and then finally 36ths.

You can do something similar to find 40ths by starting with 5ths. To do this, you need to first visualize the ¼ mark and approximate the ⅓ mark from that (Figures 2.31.20–2.31.21).

Using the Fujimoro Approximation, you can divide into 5ths.

Figure 2.31.22 shows the 10ths, 20ths, and finally the 40ths grid. What other grid densities can you find using this technique?

2.32 "Front" and "Reverse" Sides

I've been calling one side of the paper the front and the other the reverse throughout most of this book. But that doesn't make much sense, because the piece can be displayed quite well on either side, and it's often difficult to determine which side is more "valuable" or interesting as a design. During folding, I use the term *front* to refer to the side of the paper where the twists lie. This is because this is the side of the paper I'm viewing more often during the folding process. Naturally, the other side is called the *reverse*.

In the next section, I'll explore what happens when you use the pleats on both sides of the paper. Before you begin the next pattern, I wish to define two terms that will be useful. When the pleats intersect and push most of the paper of the molecule toward the viewer, we will call those pleats *top pleats*. If they push most of the paper away from the viewer, we will call them *bottom pleats*. These terms are used when the entirety of the molecule is pushed on one side of the paper or the other, and not when there is a mix. These will be used to help us fold the tendril tessellations.

2.33 Tendril Tessellation

You'll start the tendril tessellation with a hexagonal backtwist. Flip the paper to show the bottom pleats of the molecule.

One at a time, reorient each of the bottom pleats of the molecule; they will not lie flat yet. Once you have all six started, press down, and there will be skewed triangle twists that begin to form in a star pattern. They will also not lie flat, but you can get them very close to flat (Figure 2.33.2).

From the tip of one of those triangle twists, count three spaces away and split the pleat as a setup for finding the next hex twist. Do this to all six bottom pleats and flip the paper to the front (Figure 2.33.3).

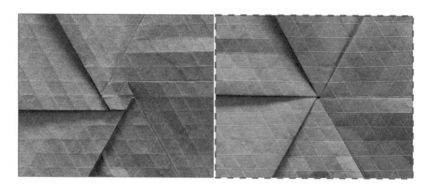

Figure 2.33.1 Hex backtwist, top pleats and bottom pleats.

Figure 2.33.2 Reorienting the bottom pleats and flattening the triangle twist.

Figure 2.33.3 Splitting to set up the triangle twists.

Unlock one of the locks into a hexagonal backtwist and flip to the reverse side again. Reorient the bottom pleats of the second hex twist into another star form (Figure 2.33.4).

Split the bottom pleats coming off the second molecule. This will result in several layers of paper piling up, but you'll resolve that soon. Flip to the front side again (Figure 2.33.5).

On the front side, there will be several locks to resolve, but fortunately you can resolve most of them at once by unlocking the closest locks to the first two molecules into a new hex backtwist. Flip the paper to the reverse again (Figure 2.33.6).

Reorient the pleats of the third hexagonal backtwist, forming a third star molecule (Figure 2.33.7). Now, the pleats between them are set

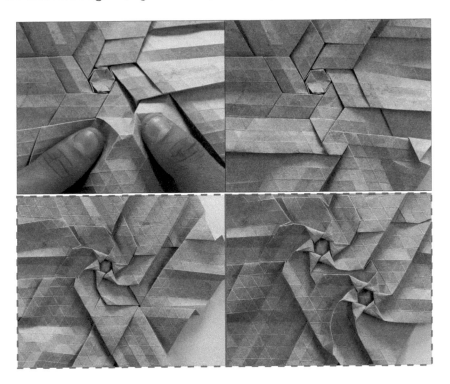

Figure 2.33.4 Unlocking into the second star.

179

Figure 2.33.5 Setting up the third star.

Figure 2.33.6 The third hex twist.

Figure 2.33.7 The third star and triangle twist.

Figure 2.33.8 Finished tendril tessellation. LEFT: Reverse. RIGHT: Front.

up for a triangle twist on the reverse of the pattern. Flatten those pleats into a triangle twist.

Continue the pattern—making a hexagonal backtwist, flipping, reorienting the bottom pleats, and creating triangle twists on the reverse when surrounded by star molecules—until the paper is tiled fully. Figure 2.33.8 shows the finished tessellation.

2.34 Inverting a Pleat

Inverting a pleat is the same as drifting and reorienting a pleat. It's a simple concept

Figure 2.34.1 Simple pleat on the diagonal.

in isolation, but it can have a drastic visual and formative effect on the two molecules connected by a pleat that becomes inverted.

To get a feel for inverting, start with a simple pleat on the diagonal of a hexagon lain flat.

Drift the pleat one grid line in the direction of the orientation of the pleat. Then reorient it. This is the finished inverted pleat. The mountain fold has become a valley fold and vice versa (Figure 2.34.2).

When connected with a molecule, the part of the platform that was on top of the pleat prior to the inversion will typically transfer below the pleat after the inversion. You'll start with inverting a pleat connected to a triangle twist. To do this, unfold the pleat you want to invert as much as you can without unfolding the twist entirely. Put your finger underneath the pleat to open it up from the back (hidden in the photos). Then, take one of the corners of the triangle and push it down through the paper. Change the parity of the rest of the pleat connected to that corner (Figure 2.34.3).

Fold the pleat and twist up again so it rests flat again. You've just inverted a pleat of the triangle twist. The second and third photos of Figure 2.34.5 show what happens when two or all three are inverted the same way. Notice that when all three pleats are inverted, it's the

181

Figure 2.34.2 Inverting the pleat.

Figure 2.34.3 Starting to invert a pleat on a triangle twist.

Figure 2.34.4 Inverting one pleat at a time.

Figure 2.34.5 Hex twist.

same as the molecule having been folded on the other side of the paper.

The hex twist offers some more variety with inverting. Apart from just inverting a single pleat, you can keep symmetry by inverting two opposite pleats or every other pleat.

If you invert opposite pleats of a hex twist and then reorient those newly-inverted pleats, it creates a fun form that looks like a split rhombic twist (Figure 2.34.6).

Inverting every other pleat of a hex twist results in a molecule that is the same on the front and reverse of the paper, which can be quite beautiful. Reorienting the inverted pleats from this structure results in the pinwheel molecule in the third photo of Figure 2.34.7.

183

Figure 2.34.6 TOP LEFT: Opposite pleats inverted. TOP RIGHT and BOTTOM ROW: Reorienting the inverted pleats.

Figure 2.34.7 LEFT: Hex twist, every other pleat inverted. MIDDLE and RIGHT: Reorienting the inverted pleats.

2.35 Iso-Area Triangle Twist Tessellation

Sometimes you are folding and come across a pattern where one side of the paper is indistinguishable from the other. This can happen with systematic pleat inversion (such as the molecule the first photo of Figure 2.34.7), but we can also flip the paper between molecules and play with that feature

Figure 2.35.1 Starting triangle twist.

Figure 2.35.2 Splitting a bottom pleat.

Figure 2.35.3 Second row of triangle twists.

Figure 2.35.4 Unlocking into the third row.

of the paper. A pattern that has the same design on both sides of the paper (with only isomorphic changes allowed between them, such as translation or rotation) is called an *iso-area tessellation*. The most basic iso-area tessellation is the 3^6 tiling, or iso-area triangle twist tessellation.

To fold this pattern, start with a triangle twist, rotated CCW in the center of the paper. Call this the front and the other side the reverse.

Flip the paper to the reverse side. Just as on the front there are three pleats, but these are bottom pleats of the first triangle twist. Split one of these bottom pleats as you would

Figure 2.35.5 Continuing the twists.

Figure 2.35.6 Finished iso-area triangle twist tessellation. LEFT: "Front". RIGHT: "Reverse."

normally to create a new triangle twist. Make it so the split happens two spaces away from the center of the paper, as shown in Figure 2.35.2. The parent pleat and its children become the top pleats of the next twist.

Then form that into a new twist. Do the same with the other two pleats on this side (Figure 2.35.3). Then flip the paper to the front and split another pleat as shown (Figure 2.35.4). You'll have to weave the bottom pleat of an eventual twist for the other side to make it work (third photo of 2.35.4). There will be a lot of flipping the paper back and forth, so prepare yourself for that!

Flip back to the reverse and notice that there is a lock between one twist's top pleat and another twist's bottom pleat. Unlock those pleats and split the bottom pleat so that the child of the split goes directly into the other twist as shown. Flip the paper back to the front to finish off that last twist in the cluster (Figure 2.35.5).

Keep working around the paper, flipping back and forth and resolving the splits until the paper has been fully tiled with twists.

Figure 2.35.6 shows the finished front and reverse sides. You can see they look exactly the same, except the front side has a twist in the center.

187

2.36 Pleat Pushing and the Platform Tessellation

One of the more interesting effects one can do with pleats is push them toward each other. When drifted so they don't intersect at a point (such as in a standing triangle spread), they can lead to some rather impressive effects!

Pleat pushing involves lifting pleats to standing position and then pushing them toward each other so that the molecule uses up more paper and the pleat becomes thicker. You can see this effect in Figure 2.36.1. Next, you'll take that skill to a full tessellation.

Start with a CCW pushed triangular spread. Next, partially unfold as shown in Figure 2.36.3 and split one of the pleats. The pleat that is being split is two triangle heights in width so the children will be two wide as well. The split will be three units away from the corner of the triangle platform (second photo of 2.36.2).

Work your way around the first cluster, creating six triangle pushes as shown in Figure 2.36.3.

Use this pushing technique to complete the entire tessellation. Notice the depth you can achieve with the pushed pleats molecules. The finished tessellation is shown in Figure 2.36.4.

Like with other molecules, with a denser tiling you can achieve a denser assortment by decreasing the spacing between the pieces. Figure 2.36.5 shows a spacing of two grid lines.

Figure 2.36.1 Pleat pushing.

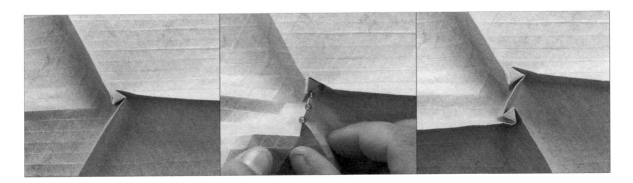

Figure 2.36.2 Forming the second molecule.

Figure 2.36.3 First cluster of molecules. RIGHT: Reverse.

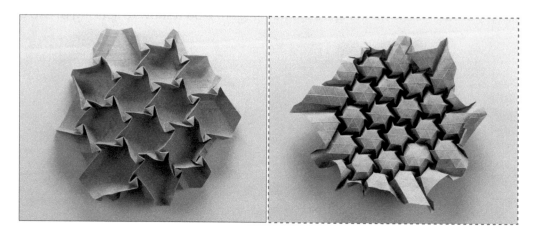

Figure 2.36.4 Finished platform tessellation. LEFT: Front. RIGHT: Reverse.

Figure 2.36.5 Alternate tiling and spacing of two.

2.37 Triple Twist Tessellation

Pleat pushing isn't just useful for a three-dimensional effect. This tessellation uses pleat pushing to create an entirely new molecule formed with composite pleats, which I call the *triple twist tessellation*.

To start, begin with a standing triangle spread as shown in Figure 2.37.1. Don't flatten the twist just yet, but create the base pattern of triangle spreads, spaced four grid lines apart (Figure 2.37.2). Again, you will scaffold this tessellation to get more complexity.

The third photo of Figure 2.37.2 shows the finished base tessellation. You have room for a few more molecules, but it's sometimes nice to leave the edges of the paper alone.

Flip the paper to the reverse. The bottom pleats of the standing spread form triangular-shaped holes. For this step, it is easier to start from the edge of the paper than the middle. Pinch pleats of two of the holes and twist CCW, creating a stack of pleats on each other (Figure 2.37.3).

Continue this pinching until every hole has a stack of pleats merging into it. The pleats will not lie flat just yet (Figure 2.37.4).

Now, start to lay the pleats flat slowly. The edges of the holes will open up, creating three triangle twists simultaneously. This is the triple triangle twist. Do this to every intersection (Figure 2.37.5).

Figure 2.37.6 shows the finished triple triangle twist tessellation.

Figure 2.37.1 Triangle spread setups and a spacing of four.

Figure 2.37.2 Finished first-tier tessellation.

Figure 2.37.3 Pinching and twisting the bottom pleats.

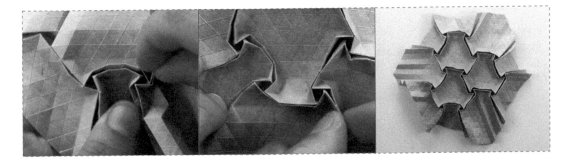

Figure 2.37.4 Finishing the second-tier tessellation.

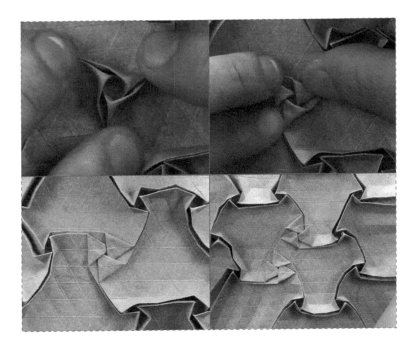

Figure 2.37.5 Flattening the triple twist molecules.

Figure 2.37.6 Finished triple twist molecule tessellation. LEFT: Reverse. RIGHT: Front.

Chapter 3

Pleat Patterns as Artwork

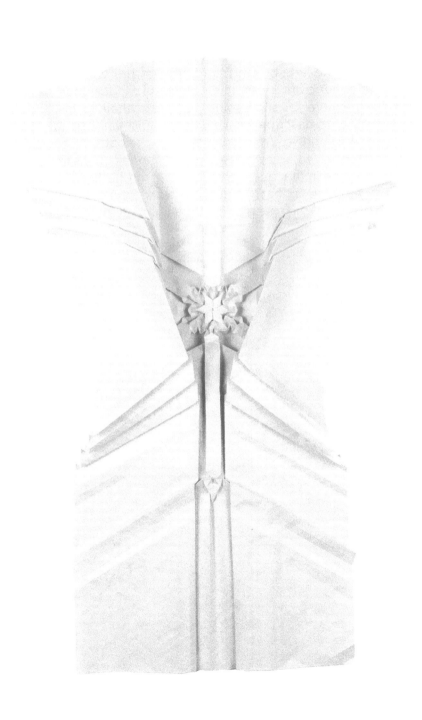

Pleat patterns are rather technical, but they also look impressive. In my time of folding them, my perception has evolved from viewing them as exercises to regarding them as an art form. This mindset has brought me to use them as an expression of the way I approach understanding the world in general.

It turns out that origami tessellations have a fairly specialized role to play in the art world. They are intriguing, but the fact that it's such a niche market leads to the artist being a constant teacher of his or her craft. I view this as an advantage, since the more you show your work, the more your ability to describe what's happening to the paper will improve. When you submit an origami tessellation to an art show, you will almost certainly hear the questions "How many sheets of paper is that?" and "Where do you buy paper with the triangles in it?" Be prepared to talk about your style, how it fits into the origami community, and its role in exploration of the medium. You should even be prepared to teach, since a lot of people will be curious about the craft!

3.1 Gallery

Alessandro Beber: Beber is an italian artist who specializes in sprawling grids of molecules. Some of the works here exhibit tilings with higher-order polygons—such as dodecagons—to create their forms. Others are extremely dense and vast triangle twist tessellations—such as the pattern shown in Section 2.3—with heavy modifications to the molecules, which he explores in his book *Origami New Worlds* [4]. His work creates beautiful shadows when held to a light.

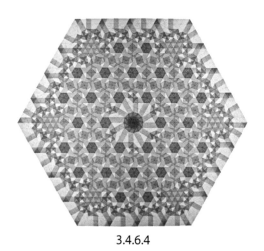

3.4.6.4

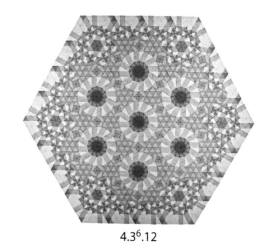

$4.3^6.12$

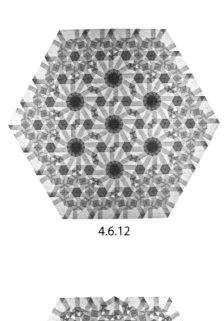

4.6.12

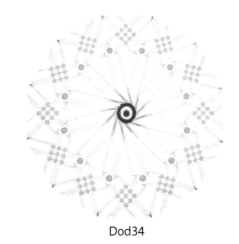

Dod34

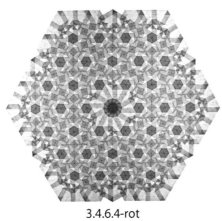

3.4.6.4-rot

Escape

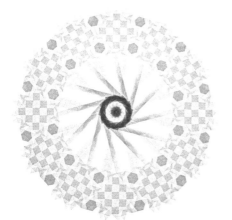

Dod3

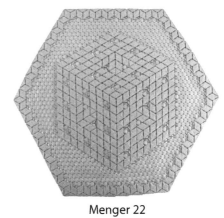

Menger 22

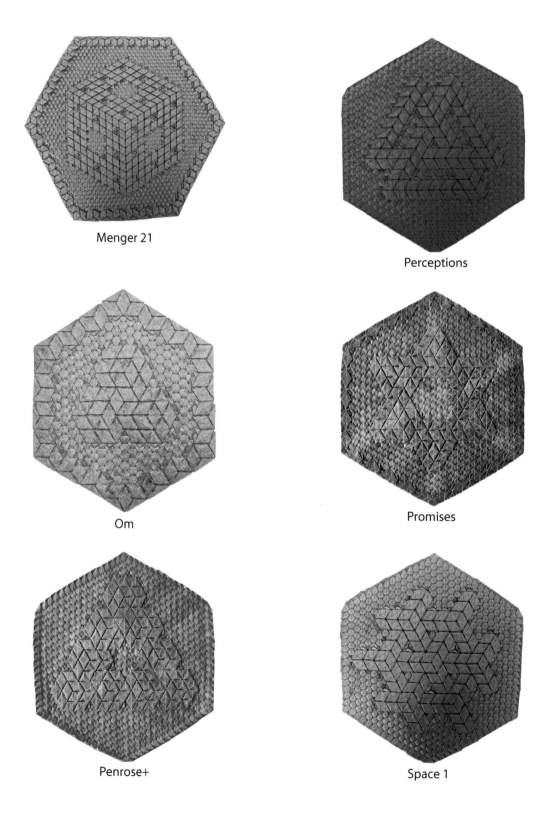

Menger 21

Perceptions

Om

Promises

Penrose+

Space 1

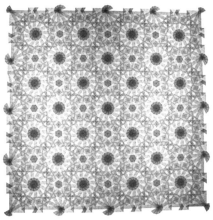

Sqdod

Joel Cooper: Most well-known for his origami masks using pleat pattern techniques, Cooper is a master of the pleated form. The piece shown here, *Medallion*, is a beautiful example of his ability to blend straight lines and curves in paper. The center is a regular hex twist surrounded by splits, and you can see how the pattern might repeat if given the room. Connecting the motifs are wavy standing pleats, a striking contrast to the platform in the center.

Sqdod 1

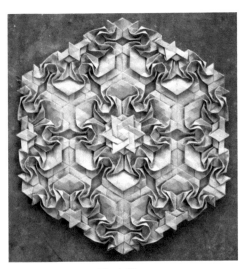

Medallion

Sqdod 2

199

Ilan Garibi: Origami artist and designer Garibi designs pleated forms for paper and metal alike. His designs are often functional, such as wearables, lighting fixtures, or furniture, and he has showcased his designs internationally. In *Hidden*, Garibi uses a square tessellation and a pleat decoration called a *puff* to create a rounded form on the reverse side of the paper. His book *Origami Tessellations for Everyone: Original Designs* features several original tessellations using the square grid [5].

Hidden (front)

Hidden (reverse)

Melina "Yureiko" Hermsen: Pleats do not have to be parallel, or even folded flat; Hermsen's pieces show this off very well. She uses a crumpling technique and methyl cellulose to have a drastic shift between the ordered nature of pleats and the chaos surrounding the molecule. Intersecting pleats displace an area of paper to form a molecule, but once the molecule is formed, one can think of the pleats as extra paper, and Hermsen makes good use of that element in her works.

Bull Riding

Fraction

Eagle

Hidden

Falling

Lionheart

Michał Kosmulski: These pieces are very well-ordered and interesting examples of different molecule studies. The piece *Union Jack* is an interesting take on spread twists as viewed from the reverse of the twist, and *Star Interlace Tessellation* uses densely-packed molecules with their overlaps folded under. Kosmulski says that his focus is more often on the molecules while pleats are mostly a by-product of the design.

Her Majesty's Tessellation

Star of David Tessellation

Star Interlace Tessellation

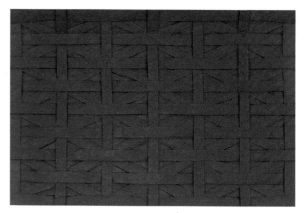

Union Jack

Robert Lang: Author of more than a dozen books on origami, including the staple opuses *Origami Design Secrets* and *Twists, Tilings, and Tessellations*, Lang is recognized as one of the world's leading masters of the art [1,6]. While primarily known for his extremely realistic origami animal folds, the pieces shown here are from his more geometric collection. I particularly enjoy the mental puzzle behind the *Hyperbolic Limit*, which uses heptagonal (seven-fold) symmetry to create its tiling.

Hyperbolic Limit

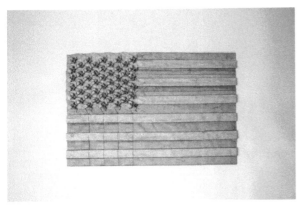

Flag

Rock Climber

Ben Parker, in collaboration with Tainted Inc.: In 2015, I was asked if I would be interested in collaborating on a dress design with a creative arts studio in Hartford. They had a vision for a dress, and we worked on getting the design to match their vision. Eventually, it was worn on the runway at *Trashion Fashion 2016* at Hartford City Hall and is still on display at the studio at Tainted Inc. The dress was created with modifications of the rhombic twist, and you can see the reverse of the tessellation in the train of the piece.

Trashion Fashion Dress

Ben Parker, in collaboration with Christine Dalenta: I was part of a multi-year collaboration with alternative photographer Christine Dalenta, during which time we created a vast series of works. Dalenta has been an alternative photographer and educator for more than forty years. She met me and shared her idea to fold light-sensitive paper and expose it while folded to record the light, modulated by the folds.

The pieces here are some of the pleat pattern works that we worked on. *Ridged Molecule Study 2* uses a hex twist with a crooked split (Section 2.27) and complicated ridgelines, while *Snowflake Flagstone with Embedded Crumple Study* is the snowflake molecule (Section 2.28) with a heavy initial light exposure before we unfolded the paper, crumpled it, and lightly exposed it again.

Ridged Molecule Study 2

Snowflake Flagstone Molecule with
Embedded Crumple Study

Halina Rościszewska-Narloch: The works by Rościszewska-Narloch display two things for the reader: the ability for origami tessellations to become modular works, and the ability for origami tessellations to appear as though they spring from the surface. The multi-unit forms of Rościszewska-Narloch's work offers a nice color contrast that one doesn't typically get with origami tessellations. *Meadow* gives the paper an ancient look that makes you question whether it's paper or stone.

Geastrum

Eryngium Maritimum

Geastrum (closeup)

Eryngium Maritimum (closeup)

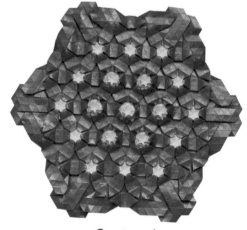

Geastrum 1

Geastrum 1 (closeup)

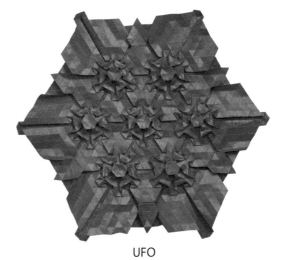

UFO

Meadow

UFO (closeup)

Meadow (closeup)

Robin Scholz: Depending on the paper used, an origami tessellation with a light coming from the back can reveal stunning interior structure; Scholz's pieces make excellent use of that feature. He uses low-density designs to allow a lot of light to shine through. The play of the light on the composite pleats between clusters of molecules are really what make piece like *Amizade, Organic vs. Technical* transcend their already complex folded patterns.

Amizade, Organic Versus Technical (reverse)

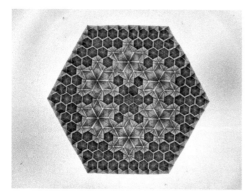

All the Stars are Shining Tonight

Helena Verrill: Approaching origami artwork from a technical standpoint, Verrill enjoys the process of designing works based on mathematical concepts from drawing based around possible ways for circles to intersect to origami pleat patterns and spread variations. This piece is an example of one of these spread twist variations, which could be interpreted as densely-packed rhombic spreads with some pleat inversions.

Amizade, Organic Versus Technical (front)

Untitled

3.2 Pleat-to-Molecule Analysis

While developing this book, two colleagues—Andrew Fisher and Matthew Benet—and I began discussing the framework of how to describe pleat patterns mathematically. Over several weeks, we developed a notation system for describing pleat intersections, introduced in Section 1.16. We will develop how we can use this in analysis in Section 3.3. The notation was our primary focus, but as our analysis grew, we were able to use it to put into context many of the other questions about the behavior of twists. Over the course of our studies, Andrew went on to analyze Brocard points (see Section 3.12), while Matt and I polished this notation system into what it is now. We then delved beyond notation and into why certain pleat intersections appeared to work better than others and developed models to explain what we observed. The remaining portions of Chapter 3 expand on Benet and my models and explanations heavily, as well as Fisher's Brocard analyses. I break the explanations into two distinct approaches: *Pleat-to-Molecule (PTM)* analysis and *Molecule-to-Pleat (MTP)* analysis. PTM analysis asks: "I have this pleat intersection. What molecules can it create?" MTP analysis asks: "I have this molecule. What pleat intersection can form it?"

Often, we are searching for molecules that are *flat-foldable*, meaning that there are not inherent conditions that prohibit the form from folding flat. You will find that certain pleat intersections cannot result in a flat-foldable molecule, and these intersections, which I will call *non-flat-foldable* pleat intersections, will be of special interest to these studies (as we search for requirements and rules for flat-foldability). From here on out, I will use the shorthand FF for flat-foldable, and NFF for non-flat-foldable.

I will start with PTM. Here I will engage in more manipulation of the previously-learned notation and add common formats of pleat intersections.

After folding some pleat intersections, you may find that some of them resist the pleats lying flat. This leads to the question of which pleat intersections are FF and which ones are not. Then I will explore the MTP approach, where I am given a twist and need to figure out the pleats that would create it. What allows a twist to flatten entirely, and which twist polygons are excluded from the "legal" choices?

3.3 Pleat Intersection Archetype Sets

As you fold more and more twists, you will likely develop an internal repertoire of molecules that you particularly enjoy folding. Maybe it's a well-known twist with an inverted or drifted pleat. Maybe it's a double-wide pleat that you flatten, and the molecule sparks something in you. A cluster of twists can act as a single unit and connect with another cluster that has corresponding pleats, allowing a significant range of patterning. These clusters can be viewed as a single molecule unit, which is why the term "molecule" is more generic than the term "twist". I find myself folding the cluster composed of a hex twist, clusters of nub twists, and the S-shape platform—the full molecule we saw in the shift rosette tessellation in Section 2.21—often just out of the pleasure of the procedure.

To understand the molecules better, it can be helpful to break down what's possible into different formats, based on the assortment of axes that contribute to their creation. I will call these formats *archetype sets*. There are seven FF archetype sets. I use a boldface font to differentiate between the various archetypes. For example, I use $\mathbb{T}$ for the triangular archetype, which stands for any molecule that only has pleats on axes $\mathcal{A}_1$, $\mathcal{A}_3$, and $\mathcal{A}_5$, or on axes $\mathcal{A}_2$, $\mathcal{A}_4$, and $\mathcal{A}_6$.

The *archetype subset* refers to the rotational orientation of the pleat intersection, showing which **specific** axes contribute to the intersection, and is denoted by a number in subscript. In other words, the difference between the subsets is a matter of your starting reference point. For example, the $\mathbb{T}$ archetype set mentioned previously is broken into two subsets, which we label $\mathbb{T}_1$ and $\mathbb{T}_2$ (Figure 3.3.1). $\mathbb{T}_1$ has pleats on axes $\mathcal{A}_1$, $\mathcal{A}_3$, and $\mathcal{A}_5$, whereas $\mathbb{T}_2$ has pleats on axes $\mathcal{A}_2$, $\mathcal{A}_4$, and $\mathcal{A}_6$. This becomes relevant when we talk about composition of pleat intersections in Section 3.5.

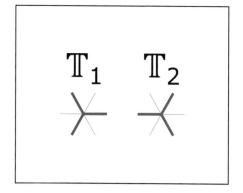

Figure 3.3.1 $\mathbb{T}$ archetype subsets.

Triangle Archetype ($\mathbb{T}$): Any molecule that uses every other axis in a hexagonal grid.

The triangle and triangle spread are two examples. An example not from the six simple twists is: $\left(\varnothing_{,1}^{0}, \varnothing_{,4}^{3}, \varnothing_{,3}^{2}, \varnothing\right)$. If the pleats come from

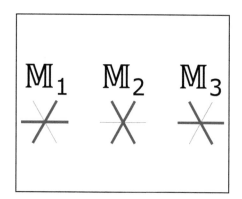

Figure 3.3.2 $\mathbb{M}$ archetype subsets.

axes $\mathcal{A}_1$, $\mathcal{A}_3$, and $\mathcal{A}_5$, it is considered a $\mathbb{T}_1$, and if the pleats come from axes $\mathcal{A}_2$, $\mathcal{A}_4$, and $\mathcal{A}_6$, it is considered a $\mathbb{T}_2$.

Rhombic Archetype ($\mathbb{M}$): Any molecule that uses four of the six axes in a hexagonal grid, where the two unrepresented axes are opposite each other.

The rhombic twist and nub molecules are two examples. An example not from the six simple twists is $\left(\varnothing_{,1}^{0}{}_{,3}^{2}, \varnothing_{,2}^{1}{}_{,1}^{0}\right)$. Remember that the numbering system notes which axis contributes to the first (or trailing) pleat of the largest cluster of consecutive pleats. Thus, an $\mathbb{M}_1$, which has pleats coming from axes $\mathcal{A}_1$, $\mathcal{A}_2$, $\mathcal{A}_4$, and $\mathcal{A}_5$, is named after the first trailing pleat moving clockwise from the right axis, in this case $\mathcal{A}_1$. Similarly, an $\mathbb{M}_2$ has pleats coming from axes $\mathcal{A}_2$, $\mathcal{A}_3$, $\mathcal{A}_5$, and $\mathcal{A}_6$. An $\mathbb{M}_3$ has pleats coming from axes $\mathcal{A}_3$, $\mathcal{A}_4$, $\mathcal{A}_6$, and $\mathcal{A}_1$.

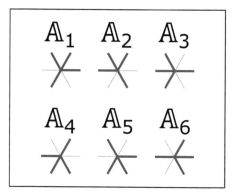

Figure 3.3.3 $\mathbb{A}$ archetype subsets.

Arrow Archetype ($\mathbb{A}$): Any molecule that uses four of the six axes in a hexagonal grid where the two unrepresented axes have exactly one axis between them.

The arrow twist is one example and the archetype's namesake. An example not from the six simple twists is $\left(\varnothing_{,1}^{0}{}_{,2}^{0}{}_{,1}^{0}, \varnothing_{,4}^{0}\right)$. An $\mathbb{A}_1$ has pleats coming from axes $\mathcal{A}_1$, $\mathcal{A}_2$, $\mathcal{A}_3$, and $\mathcal{A}_5$;

an $\mathbb{A}_2$ has pleats coming from axes $\mathcal{A}_2$, $\mathcal{A}_3$, $\mathcal{A}_4$, and $\mathcal{A}_6$; an $\mathbb{A}_3$ has pleats coming from axes $\mathcal{A}_3$, $\mathcal{A}_4$, $\mathcal{A}_5$, and $\mathcal{A}_1$; an $\mathbb{A}_4$ has pleats coming from axes $\mathcal{A}_4$, $\mathcal{A}_5$, $\mathcal{A}_6$, and $\mathcal{A}_2$; an $\mathbb{A}_5$ has pleats coming from axes $\mathcal{A}_5$, $\mathcal{A}_6$, $\mathcal{A}_1$, and $\mathcal{A}_3$; and an $\mathbb{A}_6$ has pleats coming from axes $\mathcal{A}_6$, $\mathcal{A}_1$, $\mathcal{A}_2$, and $\mathcal{A}_4$.

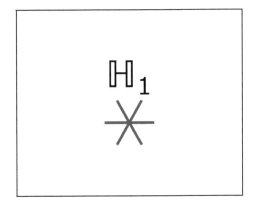

Figure 3.3.4 $\mathbb{H}$ archetype subset.

Hexagonal Archetype $(\mathbb{H})$: Any molecule that uses all six axes in a hexagonal grid.

The hex and hex spread twists are two examples. An example not from the six simple twists is $\left(\begin{smallmatrix}0&0&0&0&0&0\\1&,2&,1&,2&,1&,2\end{smallmatrix}\right)$. The only subset is $\mathbb{H}_1$, with pleats coming from every axis.

However, the $\mathbb{T}$, $\mathbb{M}$, $\mathbb{A}$, and $\mathbb{H}$ are not the only possibilities. For example, not included in the six simple twists are the arrangements with pleats coming from five axes, two axes, or none at all.

Pentagonal Archetype $(\mathbb{P})$: Molecules that use five out of six of the axes.

An example is $\left(\emptyset,\begin{smallmatrix}1&0&0&0&0\\,2&,1&,1&,1&,2\end{smallmatrix}\right)$. A $\mathbb{P}_1$ has pleats coming from axes $\mathcal{A}_1$, $\mathcal{A}_2$, $\mathcal{A}_3$, $\mathcal{A}_4$ and $\mathcal{A}_5$; a $\mathbb{P}_2$ has pleats coming from axes $\mathcal{A}_2$, $\mathcal{A}_3$, $\mathcal{A}_4$, $\mathcal{A}_5$ and $\mathcal{A}_6$; a $\mathbb{P}_3$ has pleats coming from axes $\mathcal{A}_3$, $\mathcal{A}_4$, $\mathcal{A}_5$, $\mathcal{A}_6$ and $\mathcal{A}_1$; a $\mathbb{P}_4$ has pleats coming from axes $\mathcal{A}_4$, $\mathcal{A}_5$, $\mathcal{A}_6$, $\mathcal{A}_1$ and $\mathcal{A}_2$; a $\mathbb{P}_5$ has pleats coming from axes $\mathcal{A}_5$, $\mathcal{A}_6$, $\mathcal{A}_1$, $\mathcal{A}_2$ and $\mathcal{A}_3$; and a $\mathbb{P}_6$ has pleats coming from axes $\mathcal{A}_6$, $\mathcal{A}_1$, $\mathcal{A}_2$, $\mathcal{A}_3$ and $\mathcal{A}_4$.

We can also have a sheet folded with just a single pleat running from one side of the paper to the other, which we will call the "bilateral molecule."

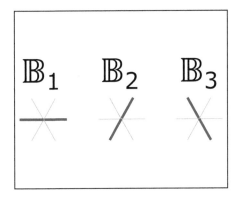

Figure 3.3.6 $\mathbb{B}$ archetype subsets.

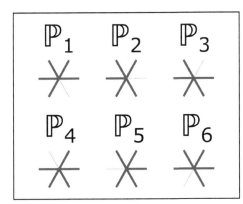

Figure 3.3.5 $\mathbb{P}$ archetype subsets.

Bilateral $(\mathbb{B})$: A straight pleat through the paper that does not intersect with another. Two opposing axes are represented in the notation, and no others. An example is $\left(\emptyset,\emptyset,\begin{smallmatrix}0\\,2\end{smallmatrix},\emptyset,\emptyset,\begin{smallmatrix}0\\,-2\end{smallmatrix}\right)$. Notice that in a bilateral, one of the pleats must be oriented in the opposite direction of the other. A $\mathbb{B}_1$ has pleats coming from axes $\mathcal{A}_1$ and $\mathcal{A}_4$. A $\mathbb{B}_2$ has pleats coming from axes $\mathcal{A}_2$

and $\mathcal{A}_5$. A $\mathbb{B}_3$ has pleats coming from axes $\mathcal{A}_3$ and $\mathcal{A}_6$.

We must also include the case where none of the axes are represented, or no folds are placed on the paper.

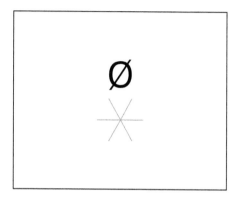

Figure 3.3.7 *Ø archetype subset.*

Null (Ø): No folds made in the paper. The only pleat intersection in this set is (Ø, Ø, Ø, Ø, Ø, Ø).

This results in seven FF molecule archetypes: Ø, $\mathbb{B}$, $\mathbb{T}$, $\mathbb{M}$, $\mathbb{A}$, $\mathbb{P}$, and $\mathbb{H}$, broken into twenty-two archetype subsets. Remember that these archetypes describe a format, an entire set of pleat intersections, and not just one pleat intersection; in fact, the number of pleat intersections in an archetype set (other than Ø) is quite vast, limited only by the number of subdivisions on your grid. The format simply describes the axes that are used to contribute

to the notation. It does not describe the distance of the mountain folds from those axes, the pleat width, or what the crease pattern of their eventual intersection is. Every one of those choices changes the molecules that might be created and has the ability to change whether a pleat intersection creates an FF molecule or not.

3.4 Molecule Database

The six simple twists are a sampling of the possible molecules. Beyond those, you learned the nub, kite, and button molecules, along with different variations of pleat splitting, and different ways to connect them on the paper. Over time, you will build a mental repertoire of more molecules, different variants of those molecules, and the combinations that work well on the paper. This database is to help you start building your own mental database for yourself. I put forth the molecules shown in the lower left of each of the photo clusters in this section as my preferred initial molecule, and any subsequent figure shows a variation of this initial molecule. The database is divided into sections based on archetypes, $\mathbb{T}$, $\mathbb{M}$, $\mathbb{A}$, $\mathbb{P}$, and $\mathbb{H}$.

$\mathbb{T}$ Molecules

The $\mathbb{T}$ molecules fit in nicely in clusters of six (as in a 3^6 pattern) or with hexagonal molecules in a 3.6.3.6 pattern.

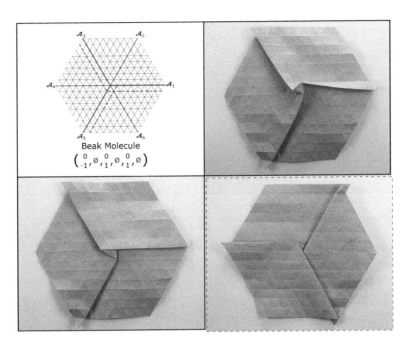

Figure 3.4.1 Beak molecule. TOP LEFT: Pleat schematic. TOP RIGHT: Standing form. LOWER LEFT: Front. LOWER RIGHT: Reverse.

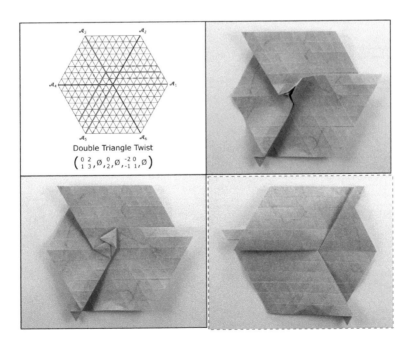

Figure 3.4.2 Double triangle twist. TOP LEFT: Pleat schematic. TOP RIGHT: Standing form. LOWER LEFT: Front. LOWER RIGHT: Reverse.

Figure 3.4.3 Double-triangle twist alternate layerings.

The beak molecule (Figure 3.4.1) is formed by reorienting a pleat of the standing form of the triangle spread twist (Figure 3.4.1).

The double-triangle twist (Figure 3.4.2) uses a combination of single-width composite pleats and one double-width simple pleat (Figure 3.4.2). The second variant from Figure 3.4.3 looks like a new pleat format for an arrow twist.

You learned about the triple triangle twist in Section 2.38. In that section you created it

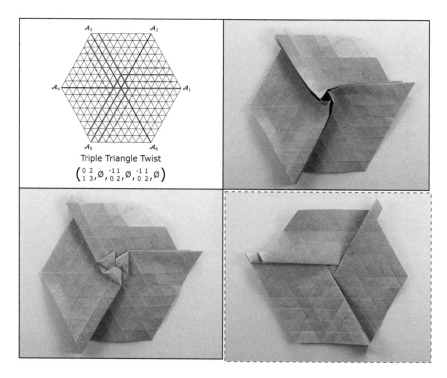

Figure 3.4.4 Triple triangle twist. TOP LEFT: Pleat schematic. TOP RIGHT: Standing form. LOWER LEFT: Front. LOWER RIGHT: Reverse.

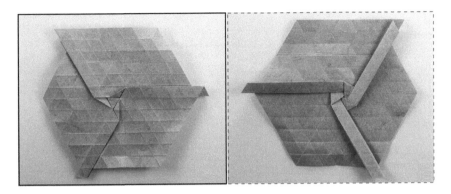

Figure 3.4.5 Triple triangle twist alternate layering. LEFT: Front. RIGHT: Reverse.

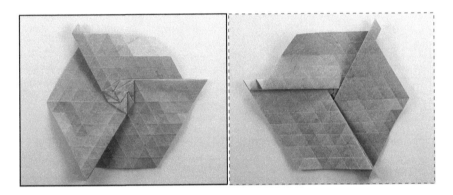

Figure 3.4.6 Another triple triangle twist alternate layering. LEFT: Front. RIGHT: Reverse.

with the standing form of a triangle spread, flipped the paper, and formed a new twist on that side. You can create the twist directly with the standing form shown in Figure 3.4.4. Rearranging the layers yields a small triangle twist on top, as in Figure 3.4.5 or 3.4.6.

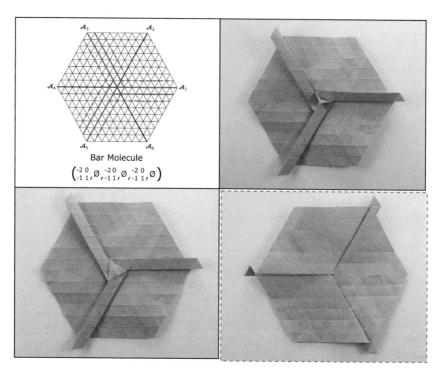

Figure 3.4.7 Bar molecule. TOP LEFT: Pleat schematic. TOP RIGHT: Standing form. LOWER LEFT: Front. LOWER RIGHT: Reverse.

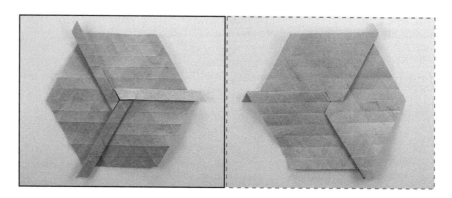

Figure 3.4.8 LEFT: Bar molecule alternate layering front. RIGHT: Bar molecule alternate layering reverse.

The bar molecule is most easily formed by starting with a triangle twist with double-wide pleats and then folding the mountain fold of the double-pleat back halfway (Figure 3.4.7). I enjoy the two variants in Figure 3.4.8 when it comes to layering.

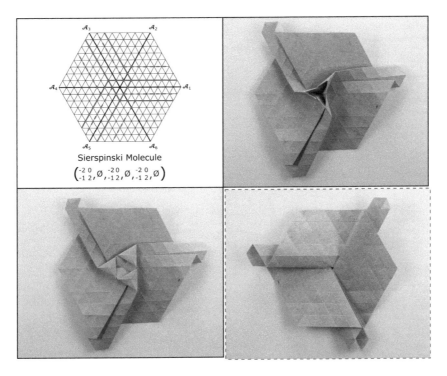

Figure 3.4.9 Sierpinski molecule. TOP LEFT: Pleat schematic. TOP RIGHT: Standing form. LOWER LEFT: Front. LOWER RIGHT: Reverse.

Folding composite pleats that look like a stair pattern can result in a stacking of three triangles connected by the corners (Figure 3.4.8).

There's a hex twist hidden inside of the triangles that you can bring to the top as well (Figure 3.4.10).

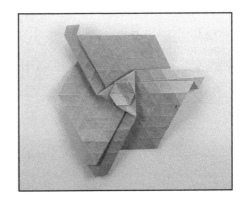

Figure 3.4.10 Sierpinski molecule alternate layering.

217

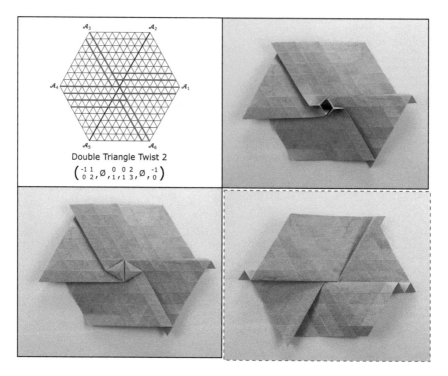

Figure 3.4.11 Double triangle twist 2. TOP LEFT: Pleat schematic. TOP RIGHT: Standing form. LOWER LEFT: Front. LOWER RIGHT: Reverse.

⋈ Molecules

Following the pattern in Figure 3.4.11, you can get a double-triangle.

There's a hidden nub in the twist that you can bring to the top as well (Figure 3.4.12).

Figure 3.4.12 Double-triangle twist alternate layering.

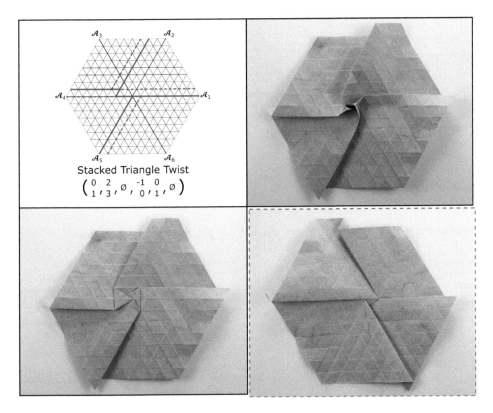

Figure 3.4.13 Stacked triangle twist. TOP LEFT: Pleat schematic. TOP RIGHT: Standing form. LOWER LEFT: Front. LOWER RIGHT: Reverse.

The stacked triangle twist (Figure 3.4.13) is another example of stacking polygons. This one has two triangle platforms, one on top of the other, at an angle to each other.

The split rhombus in Figure 3.4.15 is made by grafting a new pleat into a rhombic twist, making that pleat composite.

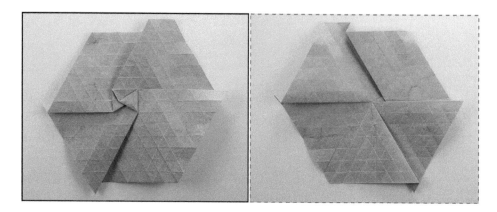

Figure 3.4.14 Stacked triangle twist alternate layering. LEFT: Front. RIGHT: Reverse.

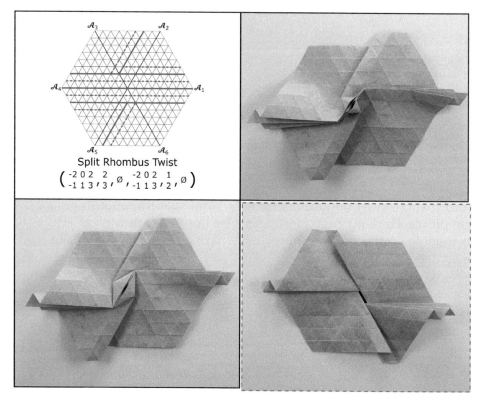

Figure 3.4.15 Split rhombic twist. TOP LEFT: Pleat schematic. TOP RIGHT: Standing form. LOWER LEFT: Front. LOWER RIGHT: Reverse.

A Molecules

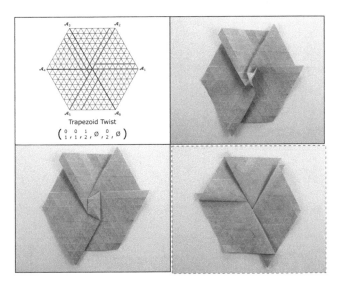

Figure 3.4.16 Trapezoid twist. TOP LEFT: Pleat schematic. TOP RIGHT: Standing form. LOWER LEFT: Front. LOWER RIGHT: Reverse.

To make the trapezoid twist (Figure 3.4.16), start with a standing CCW arrow twist and drift the leading pleat of the cluster one grid line CCW.

For the half arrow twist (Figure 3.4.17), take an arrow twist and reorient the trailing pleat of the cluster. Some reordering of layers gets this interesting molecule.

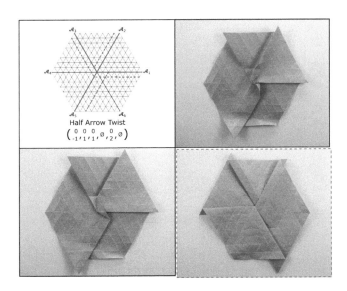

Figure 3.4.17 Half arrow twist. TOP LEFT: Pleat schematic. TOP RIGHT: Standing form. LOWER LEFT: Front. LOWER RIGHT: Reverse.

Figure 3.4.18 Half arrow twist alternate layering.

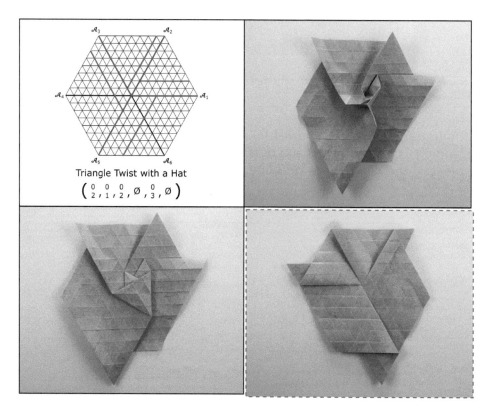

Figure 3.4.19 Triangle twist with a hat. TOP LEFT: Pleat schematic. TOP RIGHT: Standing form. LOWER LEFT: Front. LOWER RIGHT: Reverse.

The triangle twist with a hat (Figure 3.4.19) is made by grafting a $\mathbb{T}_1$ onto a standard arrow twist. This is reflected in the notation that the pleats on $\mathcal{A}_1$, $\mathcal{A}_3$, and $\mathcal{A}_5$ are one grid line wider than a normal arrow twist.

Figure 3.4.20 Triangle twist with a hat alternate layering.

ℙ Molecules

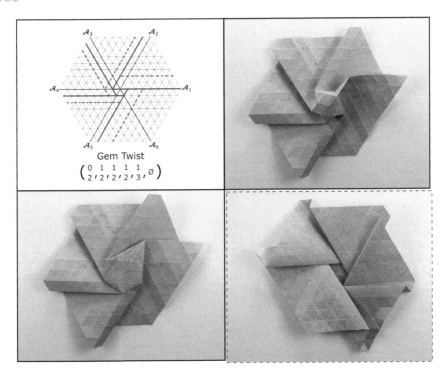

Figure 3.4.21 Gem twist. TOP LEFT: Pleat schematic. TOP RIGHT: Standing form. LOWER LEFT: Front. LOWER RIGHT: Reverse.

The gem twist (Figure 3.4.21) works out quite nicely on the paper and gives an interesting trapezoidal negative space.

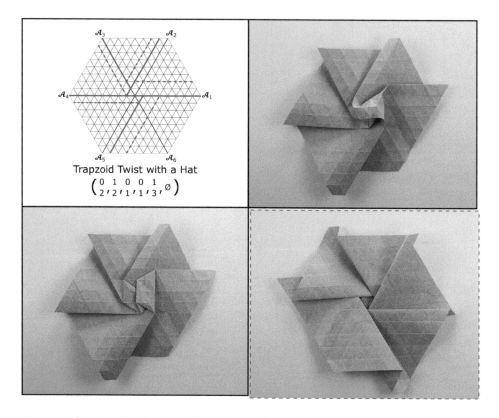

Figure 3.4.22 Trapezoid twist with a hat. TOP LEFT: Pleat schematic. TOP RIGHT: Standing form. LOWER LEFT: Front. LOWER RIGHT: Reverse.

The trapezoid twist with a hat is one way to lay this arrangement flat. Notice the interesting effect in Figure 3.4.22, resulting in both a rhombus platform and a trapezoid platform.

The snail twist (Figure 3.4.24) has a very intense notch that produces an uncommon reverse side.

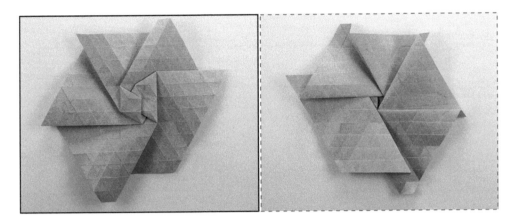

Figure 3.4.23 Trapezoid twist with a hat alternate layering. LEFT: Front. RIGHT: Reverse.

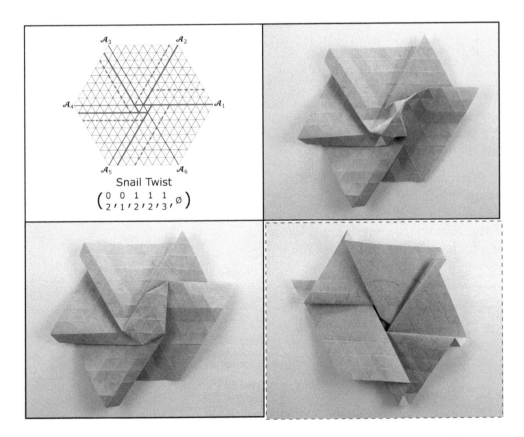

Figure 3.4.24 Snail twist. TOP LEFT: Pleat schematic. TOP RIGHT: Standing form. LOWER LEFT: Front. LOWER RIGHT: Reverse.

⊞ Molecules

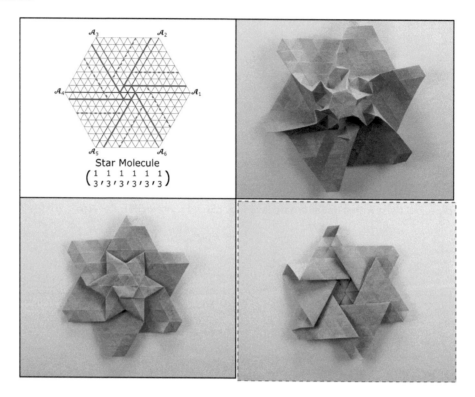

Figure 3.4.25 Star molecule. TOP LEFT: Pleat schematic. TOP RIGHT: Standing form. LOWER LEFT: Front. LOWER RIGHT: Reverse.

The star molecule (Figure 3.4.25) is a way to incorporate concavity into your designs, and the double-wide simple pleats make tiling relatively easy. Additionally, there are a lot of variants for a folder to play with.

Figure 3.4.26 Star molecule alternate layering. LEFT: Standing form. MIDDLE: Front. RIGHT: Reverse.

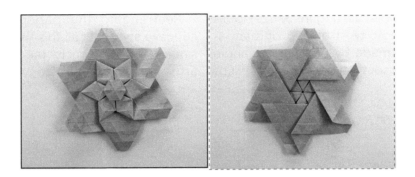

Figure 3.4.27 Star molecule alternate layering. LEFT: Front. RIGHT: Reverse.

Figure 3.4.28 Star molecule alternate layering.

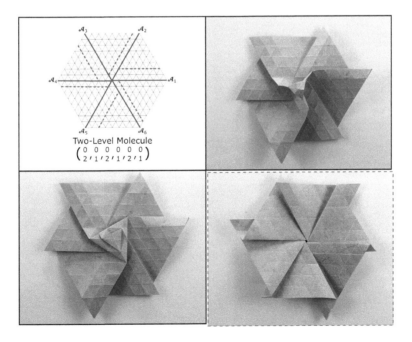

Figure 3.4.29 Two-level triangle molecule. TOP LEFT: Pleat schematic. TOP RIGHT: Standing form. LOWER LEFT: Front. LOWER RIGHT: Reverse.

The two-level molecule (Figure 3.4.29) looks like there's a triangle twist on top of a larger triangle twist. It's made by grafting a triangle twist's pleat intersection onto a hex twist's pleat intersection.

Rearranging the paper a bit gives this concave shape. I call this the *Konradical twist*, since it was shown to me by fellow origami artist Ralf Konrad at a convention some years ago.

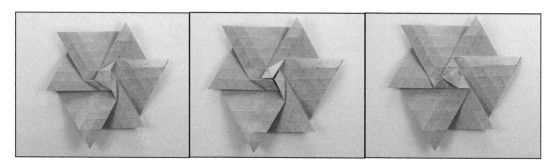

Figure 3.4.30 Two-level molecule alternate layerings. LEFT: Konradical twist.

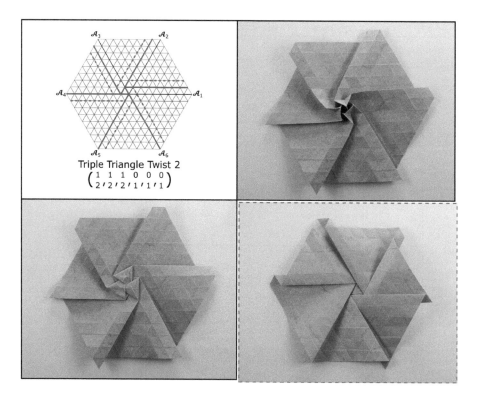

Figure 3.4.31 Triple triangle twist 2. TOP LEFT: Pleat schematic. TOP RIGHT: Standing form. LOWER LEFT: Front. LOWER RIGHT: Reverse.

Clusters of triangle twists come up often; Figure 3.4.31 is another example. This one has the same pleat pattern as the hex twist, except half of the pleats are drifted positively by one grid line.

Figure 3.4.32 Triple triangle twist 2 alternate layering.

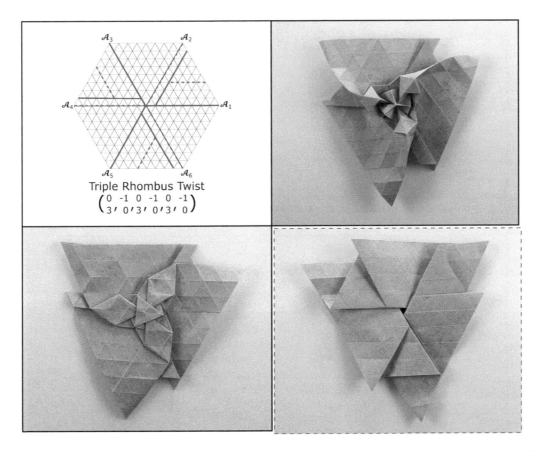

Figure 3.4.33 Triple rhombus twist. TOP LEFT: Pleat schematic. TOP RIGHT: Standing form. LOWER LEFT: Front. LOWER RIGHT: Reverse.

This interesting twist in Figure 3.4.33 has three rhombi lined up tip to tip. This is done by grafting a double-wide triangle twist onto a hex twist.

Figure 3.4.34 Triple rhombus twist alternate layerings.

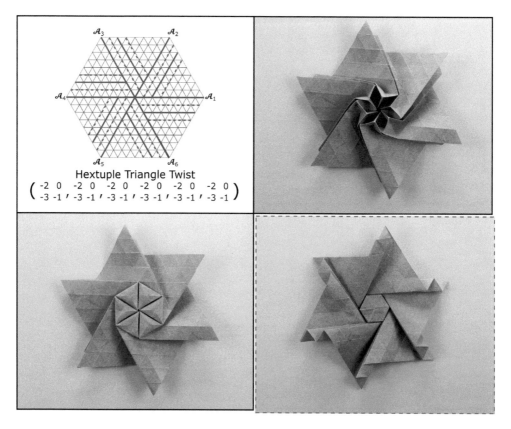

Figure 3.4.35 Hextuple triangle twist. TOP LEFT: Pleat schematic. TOP RIGHT: Standing form. LOWER LEFT: Front. LOWER RIGHT: Reverse.

The composite pleat intersection in the *hextuple triangle twist* in Figure 3.4.21 looks complicated, but it's really just an accordion fold along each axis. The negative-most mountain folds are two spaces negative of their axis, resulting in a modified hex spread.

I really enjoy this piece in standing form, and sometimes will make an entire tessellation just out of that form.

After flattening, you can also rearrange some layers to reveal the hex twist in the center (Figure 3.4.36).

Figure 3.4.36 Hextuple triangle twist alternate layerings.

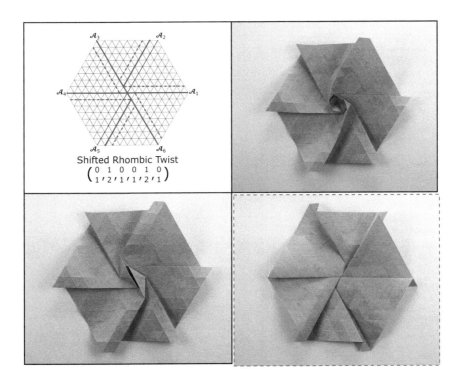

Figure 3.4.37 Shifted rhombic twist. TOP LEFT: Pleat schematic. TOP RIGHT: Standing form. LOWER LEFT: Front. LOWER RIGHT: Reverse.

Figure 3.4.38 Shifted rhombic twist alternate layering.

The shifted rhombic twist (Figure 3.4.37) creates a curious split in our familiar rhombic platform.

The pinwheel twist was mentioned in Section 2.34 and is quite fun to peel apart and lay flat.

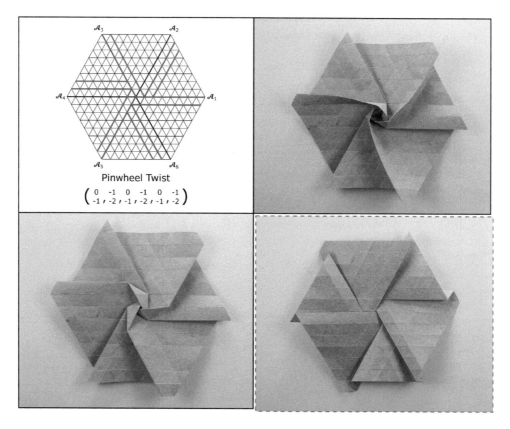

Figure 3.4.39 Pinwheel twist. TOP LEFT: Pleat schematic. TOP RIGHT: Standing form. LOWER LEFT: Front. LOWER RIGHT: Reverse.

Figure 3.4.40 Pinwheel twist alternate layering.

3.5 Archetype Composition

Several of the molecules in the database involved grafting pleats or pleat intersections onto other pleat intersections. But what exactly does it mean to graft pleat intersections together? When pleats intersect, they displace some area of paper, which I call the molecule. Generally, I've chosen to displace the paper upward toward myself and away from the table. Once the displaced paper is in a form I can manage—the standing form—I figure out how to lay it flat again.

I will call grafting one pleat intersection onto another *composition* and analyze them based on the archetype changes that occur when composing certain archetypes with other archetypes. For example, any pleat intersection that is composed with a $\mathbb{T}_1$ gains pleats on axes $\mathcal{A}_1$, $\mathcal{A}_3$, and $\mathcal{A}_5$, since those are the pleats that contribute to that archetype subset. If the initial pleat intersection already has a pleat on each of those axes, it doesn't remove it; it just makes the pleat already there wider, or it creates a composite pleat. However, if the initial pleat intersection does not have a pleat from one of those axes, its archetype will change to reflect this new axis contribution.

In any pleat intersection composed with a Ø, the result is the original pleat intersection. On the opposite end, any pleat intersection composed with an $\mathbb{H}$ pleat intersection becomes an $\mathbb{H}$ pleat intersection, as it is being composed with an intersection that already has the maximum number of pleats that can be represented on the triangle grid.

I will start by looking at the standard triangle twist—the quintessential $\mathbb{T}$ molecule—composed with a pleat running through its center, a $\mathbb{B}$ archetype; this would result in an arrow twist. Figure 3.5.1 is what composing a $\mathbb{T}$ with a $\mathbb{B}$ could look like. The red rectangles show notation changes after the composition.

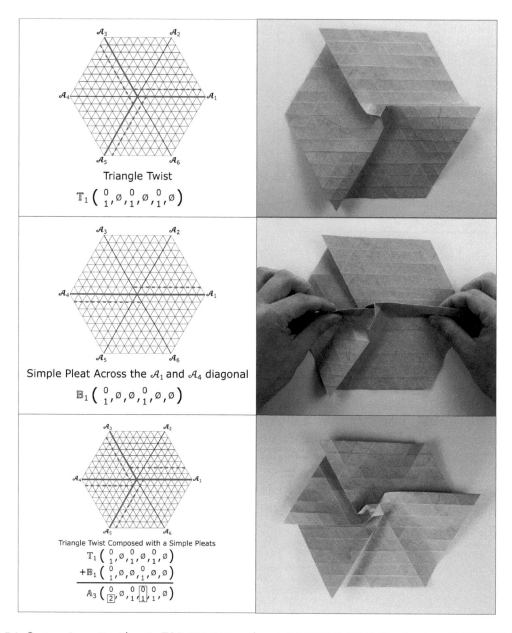

Figure 3.5.1 Composing a $\mathbb{T}_1$ with a $\mathbb{B}_1$. TOP ROW: Triangle twist. MIDDLE ROW: Simple pleat. BOTTOM ROW: Composition.

It results in an $\mathbb{A}$ molecule—in this case an $\mathbb{A}_3$. You'll also find that any $\mathbb{T}$ molecule composed with a $\mathbb{B}$ molecule will result in an $\mathbb{A}$ no matter which $\mathbb{B}$ or $\mathbb{T}$ subset was used.

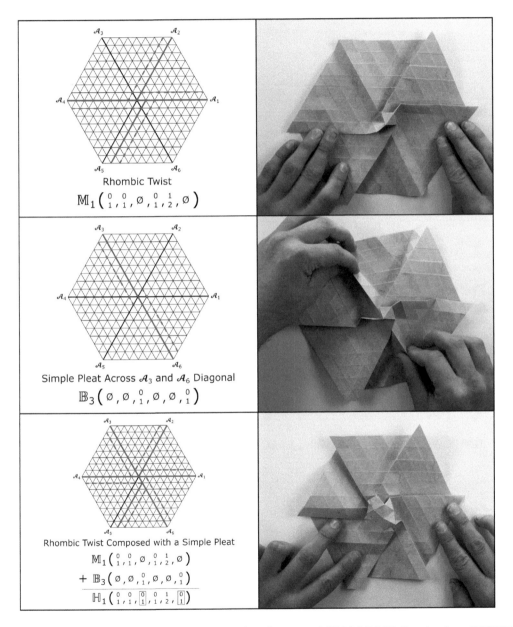

Figure 3.5.2 Composing an $\mathbb{M}_1$ with a $\mathbb{B}_3$. TOP ROW: Rhombic twist. MIDDLE ROW: Simple pleat. BOTTOM ROW: Composition.

Sometimes the subset matters when determining what the resultant archetype will be. The example in Figure 3.5.2 shows that composing an $\mathbb{M}_1$ with a $\mathbb{B}_3$ creates an $\mathbb{H}_1$, whereas composing an $\mathbb{M}_1$ with a $\mathbb{B}_1$ or $\mathbb{B}_2$ creates a wider-pleated $\mathbb{M}_1$ instead.

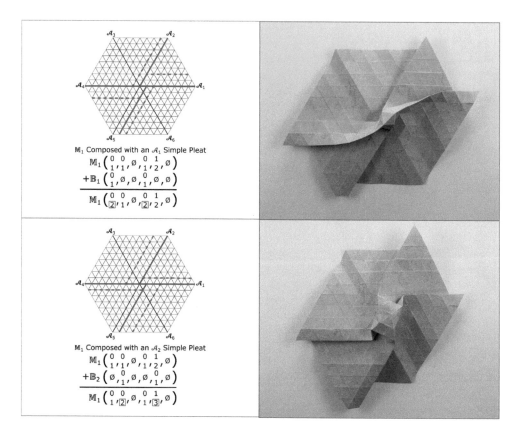

$\mathbb{M}_1$ Composed with an $\mathscr{A}_1$ Simple Pleat
$$\mathbb{M}_1 \begin{pmatrix} 0 & 0 & & 0 & 1 & \\ 1 & 1 & \varnothing & 1 & 2 & \varnothing \end{pmatrix}$$
$$+\mathbb{B}_1 \begin{pmatrix} 0 & & & 0 & & \\ 1 & \varnothing & \varnothing & 1 & \varnothing & \varnothing \end{pmatrix}$$
$$\overline{\mathbb{M}_1 \begin{pmatrix} 0 & 0 & & 0 & 1 & \\ \boxed{2} & 1 & \varnothing & \boxed{2} & 2 & \varnothing \end{pmatrix}}$$

$\mathbb{M}_1$ Composed with an $\mathscr{A}_2$ Simple Pleat
$$\mathbb{M}_1 \begin{pmatrix} 0 & 0 & & 0 & 1 & \\ 1 & 1 & \varnothing & 1 & 2 & \varnothing \end{pmatrix}$$
$$+\mathbb{B}_2 \begin{pmatrix} & 0 & & & 0 & \\ \varnothing & 1 & \varnothing & \varnothing & 1 & \varnothing \end{pmatrix}$$
$$\overline{\mathbb{M}_1 \begin{pmatrix} 0 & 0 & & 0 & 1 & \\ 1 & \boxed{2} & \varnothing & 1 & \boxed{3} & \varnothing \end{pmatrix}}$$

Figure 3.5.3 **TOP ROW:** Composing an $\mathbb{M}_1$ with a $\mathbb{B}_1$. **BOTTOM ROW:** Composing an $\mathbb{M}_1$ with a $\mathbb{B}_2$.

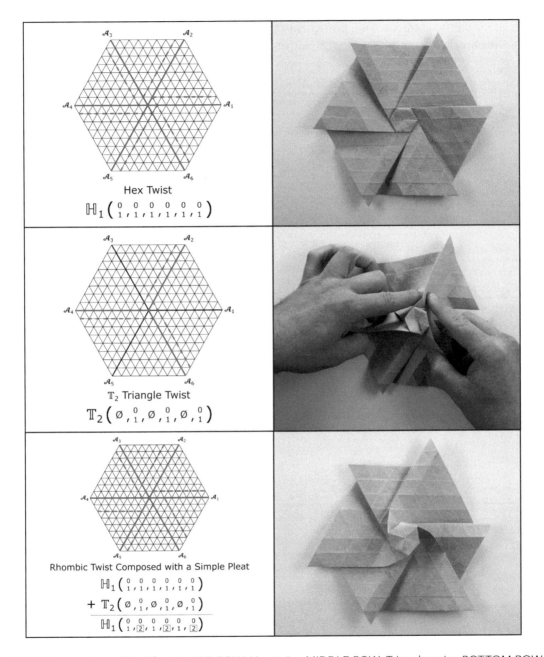

Figure 3.5.4 Composing an $\mathbb{H}_1$ with a $\mathbb{T}_2$. TOP ROW: Hex twist. MIDDLE ROW: Triangle twist. BOTTOM ROW: Composition.

Figure 3.5.4 demonstrates that composing anything with an $\mathbb{H}$ results in an $\mathbb{H}$ with wider pleats.

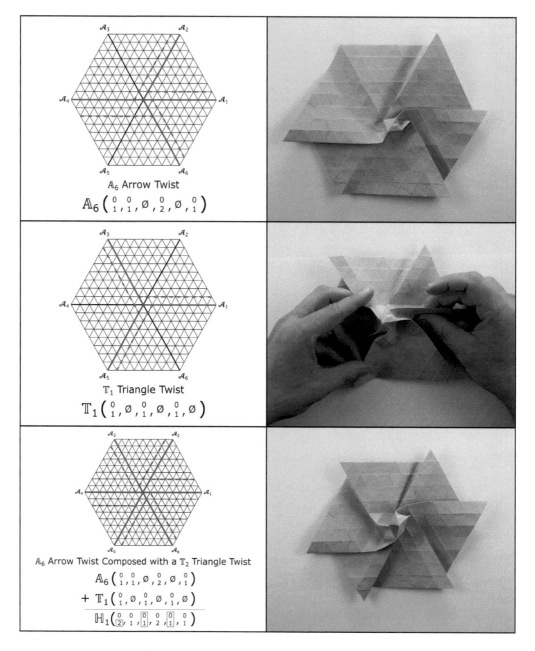

Figure 3.5.5 Composing an $\mathbb{A}_6$ with a $\mathbb{T}_2$. TOP ROW: Arrow twist. MIDDLE ROW: Triangle twist. BOTTOM ROW: Composition.

An $\mathbb{A}_6$ composed with the correctly-rotated $\mathbb{T}$ molecule results in an $\mathbb{A}_6$ with a wider tail, trailing, and leading pleat.

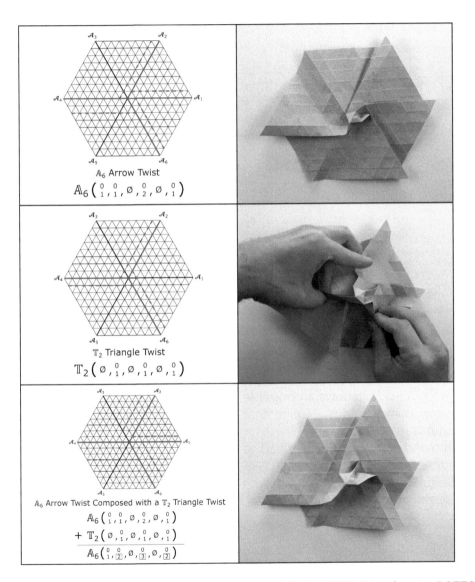

$\mathbb{A}_6$ Arrow Twist
$$\mathbb{A}_6 \left(\begin{smallmatrix} 0 \\ 1 \end{smallmatrix}, \begin{smallmatrix} 0 \\ 1 \end{smallmatrix}, \varnothing, \begin{smallmatrix} 0 \\ 2 \end{smallmatrix}, \varnothing, \begin{smallmatrix} 0 \\ 1 \end{smallmatrix} \right)$$

$\mathbb{T}_2$ Triangle Twist
$$\mathbb{T}_2 \left(\varnothing, \begin{smallmatrix} 0 \\ 1 \end{smallmatrix}, \varnothing, \begin{smallmatrix} 0 \\ 1 \end{smallmatrix}, \varnothing, \begin{smallmatrix} 0 \\ 1 \end{smallmatrix} \right)$$

$\mathbb{A}_6$ Arrow Twist Composed with a $\mathbb{T}_2$ Triangle Twist
$$\mathbb{A}_6 \left(\begin{smallmatrix} 0 \\ 1 \end{smallmatrix}, \begin{smallmatrix} 0 \\ 1 \end{smallmatrix}, \varnothing, \begin{smallmatrix} 0 \\ 2 \end{smallmatrix}, \varnothing, \begin{smallmatrix} 0 \\ 1 \end{smallmatrix} \right)$$
$$+ \mathbb{T}_2 \left(\varnothing, \begin{smallmatrix} 0 \\ 1 \end{smallmatrix}, \varnothing, \begin{smallmatrix} 0 \\ 1 \end{smallmatrix}, \varnothing, \begin{smallmatrix} 0 \\ 1 \end{smallmatrix} \right)$$
$$\overline{\mathbb{A}_6 \left(\begin{smallmatrix} 0 \\ 1 \end{smallmatrix}, \begin{smallmatrix} 0 \\ \boxed{2} \end{smallmatrix}, \varnothing, \begin{smallmatrix} 0 \\ \boxed{3} \end{smallmatrix}, \varnothing, \begin{smallmatrix} 0 \\ \boxed{2} \end{smallmatrix} \right)}$$

Figure 3.5.6 Composing an $\mathbb{A}_6$ with a $\mathbb{T}_1$. TOP ROW: Arrow twist. MIDDLE ROW: Triangle twist. BOTTOM ROW: Composition.

An $\mathbb{A}_6$ composed with a $\mathbb{T}_1$ results in an $\mathbb{H}_1$ with a wide pleat two opposing axes.

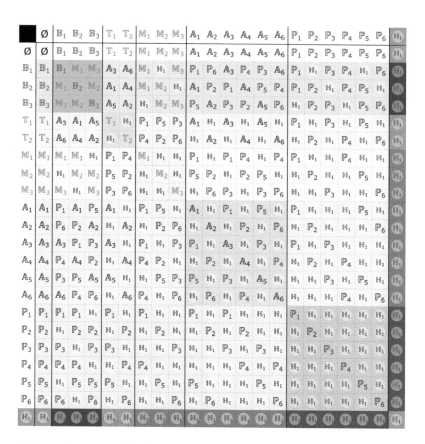

Figure 3.5.7 All possible combinations of archetype composition.

The chart in Figure 3.5.7 shows all possible pleat intersection compositions. Notice that many compositions result in H_1's.

Playing with these combinations can lead to a simply incredible number of options for pleat intersection design. I encourage you to play with a grid and to see what molecules you can get.

3.6 Actions and Notations

As you develop your intuition with Benet notation, you may ask yourself what happens as you apply different actions, such as drifting or inverting. You've seen what pleat intersection composition looks like. This section deals with how to interpret other actions such as reorientation, drifting, inverting, and splitting.

In order to do that, I need to go over two distinct forms of the notation: *absolute form* and *relative form*. The version of Benet notation I described in Section 1.16 was presented in absolute form, which describes the exact location of the mountain and valley folds on an axis. For example, $\frac{1}{2}$ describes a pleat where the mountain fold is one grid line positive of the axis and the valley fold is two grid lines positive of the axis. In relative form, we place the top left crease of the matrix—the first mountain fold counting positively—on the axis itself, modifying the numbers of the other creases to match. Then we move the amount of the decrease outside of the matrix notation as a sort of coefficient for the matrix, indicating the drift of the pleat. The same pleat noted previously looks like this in relative form: 1_1^0. The two notations are the same, just different forms. If the first mountain fold of a pleat is on the axis—meaning the top left number of the notation is 0—the two forms are have no notation difference. In other words, 0_1^0 can be written more simply as $_1^0$. No information is lost here, as distributing the 0 changes nothing in the notational matrix. The shorthand relative form looks exactly the same as absolute form because, in this case, the pleats are the same.

To convert from relative to absolute you have to distribute the coefficient into the pleat matrix; in this notation, you are adding the coefficient being distributed to each crease. To convert from absolute to relative, add or subtract from the smallest value mountain fold of the pleat until it becomes 0. Then, add or subtract the same value from each of the other numbers in the crease notation by the same amount. Finally, write the added/reduced value as a coefficient before the pleat matrix. This signifies that the pleat matrix is to be drifted (positive or negative) that amount.

$$\tfrac{3}{4} = 3\,_1^0 \qquad \tfrac{-1}{0} = -1\,_1^0 \qquad \tfrac{5}{2} = 5\,_{-3}^0 \qquad \tfrac{1}{4}\tfrac{2}{5} = 1\,_3^0\tfrac{1}{4}$$

Here are several examples of equal pleat intersections. Notice that in the last example, the coefficient of 1 is distributed into both pieces of the composite pleat.

Absolute form shows exactly where on the coordinate system the creases of the pleat lie. Relative form shows the shape of the pleat separate from its location; its location is defined by the coefficient. With those learned, you can explore different actions, most of which can only work if the notation is in either absolute or relative form.

Reorientation

This is the most straightforward action you can take to a pleat. To reorient the pleat in the notation, first put the pleat into relative form, and change the sign of the ι-value in $\tfrac{m}{v}$ format, from negative to positive or vice versa.

Figure 3.6.1 shows an example of reorienting a pleat.

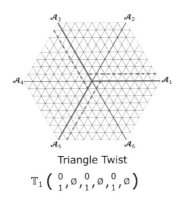

Triangle Twist

$$\mathbb{T}_1\left(\,_1^0,\varnothing,\,_1^0,\varnothing,\,_1^0,\varnothing\right)$$

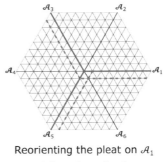

Reorienting the pleat on $\mathcal{A}_1$

$$\mathbb{T}_1\left(\,_{-1}^0,\varnothing,\,_1^0,\varnothing,\,_1^0,\varnothing\right)$$

Figure 3.6.1 Notation for reorienting one pleat.

As the notation needs to be in relative form to reorient, when the mountain fold is not on the axis, there is an extra step to convert it, reorient, and then convert back to absolute form as shown in Figure 3.6.2.

Reorientation can be called a *unary* operation, which means that there are no values to apply. This is opposed to a *binary* operation, such as integer addition, where there are two values being computed. You don't reorient a

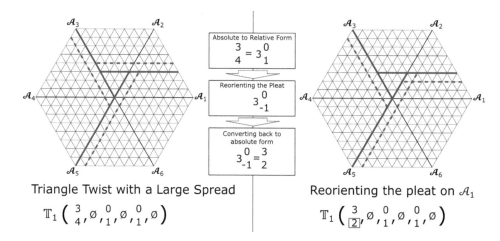

Triangle Twist with a Large Spread

$$\mathbb{T}_1\left(\begin{smallmatrix}3\\4\end{smallmatrix}, \varnothing, \begin{smallmatrix}0\\1\end{smallmatrix}, \varnothing, \begin{smallmatrix}0\\1\end{smallmatrix}, \varnothing\right)$$

Reorienting the pleat on $\mathcal{A}_1$

$$\mathbb{T}_1\left(\boxed{\begin{smallmatrix}3\\2\end{smallmatrix}}, \varnothing, \begin{smallmatrix}0\\1\end{smallmatrix}, \varnothing, \begin{smallmatrix}0\\1\end{smallmatrix}, \varnothing\right)$$

Figure 3.6.2 Notation for reorienting one pleat when the mountain fold is not on the axis.

pleat a certain amount or value. It's either in one orientation or the other. From one pleat that is lain flat, there is only one way that it can be reoriented.

Drifting

Drifting a pleat translates the mountain and valley folds of a pleat one or more grid lines in either the positive or negative direction. This does not change the archetype of the pleat intersection, and the only notation change is the coefficient in relative form increases or decreases by the number of grid lines moved. For example, the difference between the triangle twist and the triangle spread is the result of a single grid line drift on one of the pleats.

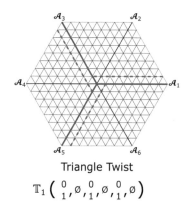

Triangle Twist

$$\mathbb{T}_1\left(\begin{smallmatrix}0\\1\end{smallmatrix}, \varnothing, \begin{smallmatrix}0\\1\end{smallmatrix}, \varnothing, \begin{smallmatrix}0\\1\end{smallmatrix}, \varnothing\right)$$

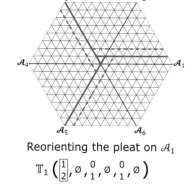

Reorienting the pleat on $\mathcal{A}_1$

$$\mathbb{T}_1\left(\boxed{\begin{smallmatrix}1\\2\end{smallmatrix}}, \varnothing, \begin{smallmatrix}0\\1\end{smallmatrix}, \varnothing, \begin{smallmatrix}0\\1\end{smallmatrix}, \varnothing\right)$$

Figure 3.6.3 Notation for drifting one pleat.

Drifting away from a twist will work to create an FF twist, but if the base twist already has a spread (such as with the rhombic twist), drifting toward the intersection makes flattening more difficult, as you'll see later. This notation shows the typical rhombic twist where you drift one pleat inward so all of the mountain folds intersect at a point.

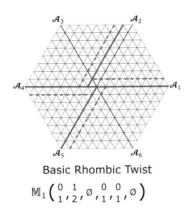

Basic Rhombic Twist

$$\mathbb{M}_1 \begin{pmatrix} 0 & 1 \\ 1 & 2 \end{pmatrix}, \emptyset, \begin{matrix} 0 & 0 \\ 1 & 1 \end{matrix}, \emptyset \end{pmatrix}$$

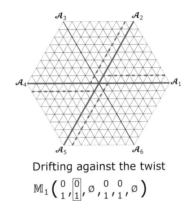

Drifting against the twist

$$\mathbb{M}_1 \begin{pmatrix} 0 & \boxed{0} \\ 1 & \boxed{1} \end{pmatrix}, \emptyset, \begin{matrix} 0 & 0 \\ 1 & 1 \end{matrix}, \emptyset \end{pmatrix}$$

Figure 3.6.4 Notation for drifting one pleat.

The notation still works and the pleats can lie flat, but the paper in the molecule butts against itself and refuses to flatten neatly. There's something going on here that's causing the paper to self-intersect. We'll explore that in Section 3.9.

Inverting

Just as reorientation is a unary action, so is pleat inversion. It consists of swapping the m and p values of absolute form. Here's an example of how the notation handles inverting a pleat.

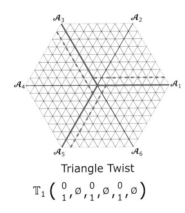

Triangle Twist

$$\mathbb{T}_1 \begin{pmatrix} 0 \\ 1 \end{pmatrix}, \emptyset, \begin{matrix} 0 \\ 1 \end{matrix}, \emptyset, \begin{matrix} 0 \\ 1 \end{matrix}, \emptyset \end{pmatrix}$$

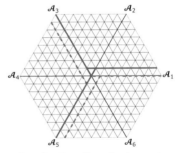

Reorienting the pleat on $\mathcal{A}_1$

$$\mathbb{T}_1 \begin{pmatrix} \boxed{1} \\ \boxed{0} \end{pmatrix}, \emptyset, \begin{matrix} 0 \\ 1 \end{matrix}, \emptyset, \begin{matrix} 0 \\ 1 \end{matrix}, \emptyset \end{pmatrix}$$

Figure 3.6.5 Notation for inverting one pleat.

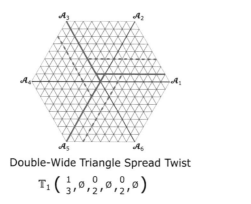
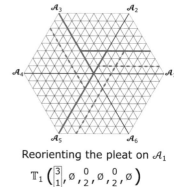

Double-Wide Triangle Spread Twist

$$\mathbb{T}_1\left(\begin{array}{c}1\\3\end{array},\varnothing,\begin{array}{c}0\\2\end{array},\varnothing,\begin{array}{c}0\\2\end{array},\varnothing\right)$$

Reorienting the pleat on $\mathcal{A}_1$

$$\mathbb{T}_1\left(\boxed{\begin{array}{c}3\\1\end{array}},\varnothing,\begin{array}{c}0\\2\end{array},\varnothing,\begin{array}{c}0\\2\end{array},\varnothing\right)$$

Figure 3.6.6 Notation for inverting one pleat.

Splitting

You saw early on in Chapter 1 that you could split a pleat on the triangle grid and create two pleats with identical width, called children pleats. In fact, this is the first step for the triangle twist. When you perform that action, the notation changes as does the archetype of the pleat intersection.

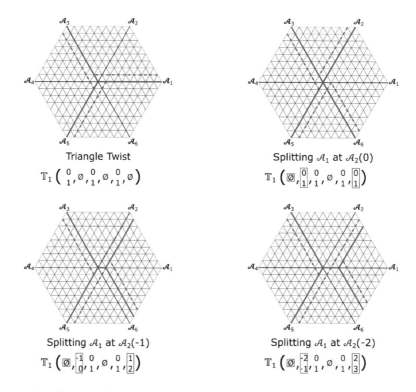

Triangle Twist

$$\mathbb{T}_1\left(\begin{array}{c}0\\1\end{array},\varnothing,\begin{array}{c}0\\1\end{array},\varnothing,\begin{array}{c}0\\1\end{array},\varnothing\right)$$

Splitting $\mathcal{A}_1$ at $\mathcal{A}_2(0)$

$$\mathbb{T}_1\left(\varnothing,\boxed{\begin{array}{c}0\\1\end{array}}\begin{array}{c}0\\1\end{array},\varnothing,\begin{array}{c}0\\1\end{array},\boxed{\begin{array}{c}0\\1\end{array}}\right)$$

Splitting $\mathcal{A}_1$ at $\mathcal{A}_2(-1)$

$$\mathbb{T}_1\left(\varnothing,\boxed{\begin{array}{c}-1\\0\end{array}}\begin{array}{c}0\\1\end{array},\varnothing,\begin{array}{c}0\\1\end{array},\boxed{\begin{array}{c}1\\2\end{array}}\right)$$

Splitting $\mathcal{A}_1$ at $\mathcal{A}_2(-2)$

$$\mathbb{T}_1\left(\varnothing,\boxed{\begin{array}{c}-2\\-1\end{array}}\begin{array}{c}0\\1\end{array},\varnothing,\begin{array}{c}0\\1\end{array},\boxed{\begin{array}{c}2\\3\end{array}}\right)$$

Figure 3.6.7 Notation for splitting a pleat. TOP RIGHT: Bar with the width of zero. BOTTOM LEFT: Bar with a width of one. BOTTOM RIGHT: Bar with a width of two.

Notice that there is a difference between splitting at the point of intersection, one grid line off, two grid lines off, etc.

Figure 3.6.8 Splitting a pleat at various bar widths.

No matter where the pleat is split in a $\mathbb{T}$, it will always result in an $\mathbb{M}$, assuming the pleat that is split reduces to nothing from the process. If the split pleat is not reduced to nothing—such as when you have a double-wide pleat that is split and contributes only one unit of its width to the surrounding axes—the result is a different archetype, as shown.

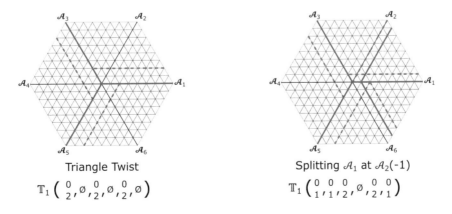

Triangle Twist

$$\mathbb{T}_1 \left(\begin{smallmatrix} 0 \\ 2 \end{smallmatrix}, \varnothing, \begin{smallmatrix} 0 \\ 2 \end{smallmatrix}, \varnothing, \begin{smallmatrix} 0 \\ 2 \end{smallmatrix}, \varnothing \right)$$

Splitting $\mathcal{A}_1$ at $\mathcal{A}_2(-1)$

$$\mathbb{T}_1 \left(\begin{smallmatrix} 0 & 0 & 0 \\ 1 & 1 & 2 \end{smallmatrix}, \varnothing, \begin{smallmatrix} 0 & 0 \\ 2 & 1 \end{smallmatrix} \right)$$

Figure 3.6.9 Notation for splitting a pleat, but not decreasing the pleat to width of zero.

Figure 3.6.10 Splitting a pleat, but not decreasing the pleat to width of zero.

Decreasing the $\mathcal{A}_1$ pleat from width of two to one means there is still a pleat coming from $\mathcal{A}_1$, which results in a $\mathbb{P}_5$ molecule, rather than an $\mathbb{M}_2$. The table in Figure 3.6.11 shows all the possible splitting combinations, the top being when the split pleat is reduced to nothing and the bottom being reduced to thinner than it was, but not completely gone.

SPLITTING TABLE (split pleat reduced to 0)

	Ø	$\mathbb{B}_1$	$\mathbb{B}_2$	$\mathbb{B}_3$	$\mathbb{T}_1$	$\mathbb{T}_2$	$\mathbb{M}_1$	$\mathbb{M}_2$	$\mathbb{M}_3$	$\mathbb{A}_1$	$\mathbb{A}_2$	$\mathbb{A}_3$	$\mathbb{A}_4$	$\mathbb{A}_5$	$\mathbb{A}_6$	$\mathbb{P}_1$	$\mathbb{P}_2$	$\mathbb{P}_3$	$\mathbb{P}_4$	$\mathbb{P}_5$	$\mathbb{P}_6$	$\mathbb{H}_1$
$\mathcal{A}_1$		$\mathbb{T}_2$			$\mathbb{M}_2$		$\mathbb{A}_4$		$\mathbb{A}_2$	$\mathbb{M}_2$		$\mathbb{P}_2$		$\mathbb{M}_2$	$\mathbb{T}_2$	$\mathbb{P}_2$		$\mathbb{P}_2$	$\mathbb{A}_4$	$\mathbb{M}_2$	$\mathbb{A}_2$	$\mathbb{P}_2$
$\mathcal{A}_2$			$\mathbb{T}_1$		$\mathbb{M}_3$		$\mathbb{A}_3$		$\mathbb{A}_5$	$\mathbb{T}_1$		$\mathbb{M}_3$		$\mathbb{P}_3$	$\mathbb{M}_3$	$\mathbb{A}_3$		$\mathbb{P}_3$	$\mathbb{P}_3$	$\mathbb{A}_5$	$\mathbb{M}_3$	$\mathbb{P}_3$
$\mathcal{A}_3$				$\mathbb{T}_2$	$\mathbb{M}_1$		$\mathbb{A}_4$		$\mathbb{A}_6$	$\mathbb{M}_1$		$\mathbb{T}_2$		$\mathbb{M}_1$	$\mathbb{P}_4$	$\mathbb{M}_1$		$\mathbb{A}_4$	$\mathbb{P}_4$	$\mathbb{P}_4$	$\mathbb{A}_6$	$\mathbb{P}_4$
$\mathcal{A}_4$		$\mathbb{T}_1$			$\mathbb{M}_2$		$\mathbb{A}_1$		$\mathbb{A}_5$	$\mathbb{M}_2$		$\mathbb{T}_1$		$\mathbb{M}_2$	$\mathbb{P}_5$	$\mathbb{A}_1$		$\mathbb{M}_2$	$\mathbb{A}_5$	$\mathbb{P}_5$	$\mathbb{P}_5$	$\mathbb{P}_5$
$\mathcal{A}_5$			$\mathbb{T}_2$		$\mathbb{M}_3$		$\mathbb{A}_6$		$\mathbb{A}_2$	$\mathbb{P}_6$		$\mathbb{M}_3$		$\mathbb{T}_2$	$\mathbb{M}_3$	$\mathbb{P}_6$		$\mathbb{A}_2$	$\mathbb{M}_3$	$\mathbb{A}_6$	$\mathbb{P}_6$	$\mathbb{P}_6$
$\mathcal{A}_6$				$\mathbb{T}_1$	$\mathbb{M}_1$		$\mathbb{A}_1$		$\mathbb{A}_3$	$\mathbb{P}_1$		$\mathbb{M}_1$		$\mathbb{T}_1$	$\mathbb{M}_1$	$\mathbb{P}_1$		$\mathbb{A}_3$	$\mathbb{M}_1$	$\mathbb{A}_1$	$\mathbb{P}_1$	$\mathbb{P}_1$

SPLITTING TABLE (split pleat reduced to >0)

	Ø	$\mathbb{B}_1$	$\mathbb{B}_2$	$\mathbb{B}_3$	$\mathbb{T}_1$	$\mathbb{T}_2$	$\mathbb{M}_1$	$\mathbb{M}_2$	$\mathbb{M}_3$	$\mathbb{A}_1$	$\mathbb{A}_2$	$\mathbb{A}_3$	$\mathbb{A}_4$	$\mathbb{A}_5$	$\mathbb{A}_6$	$\mathbb{P}_1$	$\mathbb{P}_2$	$\mathbb{P}_3$	$\mathbb{P}_4$	$\mathbb{P}_5$	$\mathbb{P}_6$	$\mathbb{H}_1$
$\mathcal{A}_1$		$\mathbb{A}_6$			$\mathbb{P}_5$		$\mathbb{P}_4$		$\mathbb{P}_6$	$\mathbb{P}_5$		$\mathbb{H}_1$		$\mathbb{P}_5$	$\mathbb{A}_6$	$\mathbb{H}_1$		$\mathbb{H}_1$	$\mathbb{P}_4$	$\mathbb{P}_5$	$\mathbb{P}_6$	$\mathbb{H}_1$
$\mathcal{A}_2$			$\mathbb{A}_1$		$\mathbb{P}_6$		$\mathbb{P}_1$		$\mathbb{P}_5$	$\mathbb{A}_1$		$\mathbb{P}_6$		$\mathbb{H}_1$	$\mathbb{P}_6$	$\mathbb{P}_1$		$\mathbb{H}_1$	$\mathbb{H}_1$	$\mathbb{P}_5$	$\mathbb{P}_6$	$\mathbb{H}_1$
$\mathcal{A}_3$				$\mathbb{A}_2$	$\mathbb{P}_1$		$\mathbb{P}_2$		$\mathbb{P}_6$	$\mathbb{P}_1$		$\mathbb{A}_2$		$\mathbb{P}_1$	$\mathbb{H}_1$	$\mathbb{P}_1$		$\mathbb{P}_2$	$\mathbb{H}_1$	$\mathbb{H}_1$	$\mathbb{P}_6$	$\mathbb{H}_1$
$\mathcal{A}_4$		$\mathbb{A}_3$			$\mathbb{P}_2$		$\mathbb{P}_1$		$\mathbb{P}_3$	$\mathbb{P}_2$		$\mathbb{A}_3$		$\mathbb{P}_2$	$\mathbb{H}_1$	$\mathbb{P}_1$		$\mathbb{P}_2$	$\mathbb{P}_3$	$\mathbb{H}_1$	$\mathbb{H}_1$	$\mathbb{H}_1$
$\mathcal{A}_5$			$\mathbb{A}_4$		$\mathbb{P}_3$		$\mathbb{P}_4$		$\mathbb{P}_2$	$\mathbb{H}_1$		$\mathbb{P}_3$		$\mathbb{A}_4$	$\mathbb{P}_3$	$\mathbb{H}_1$		$\mathbb{P}_2$	$\mathbb{P}_3$	$\mathbb{P}_4$	$\mathbb{H}_1$	$\mathbb{H}_1$
$\mathcal{A}_6$				$\mathbb{A}_5$	$\mathbb{P}_4$		$\mathbb{P}_5$		$\mathbb{P}_3$	$\mathbb{H}_1$		$\mathbb{P}_4$		$\mathbb{A}_5$	$\mathbb{P}_4$	$\mathbb{H}_1$		$\mathbb{P}_3$	$\mathbb{P}_4$	$\mathbb{P}_5$	$\mathbb{H}_1$	$\mathbb{H}_1$

Figure 3.6.11 All possible combinations of splitting.

3.7 Splitting Equation

How does splitting actually work? How can you take a single pleat and seemingly double it without the paper losing its FF state? To explore this, it is helpful to know how to create the twist without the grid. I will call folding pleat intersections without the grid *free-folding*; it is a useful skill for analyzing pleats. From there, you can generalize the twist process and design your own externally of the grid structure. For experimentation purposes, I recommend using copy paper or something cheap and strong and not being afraid to draw on the paper and mark observations.

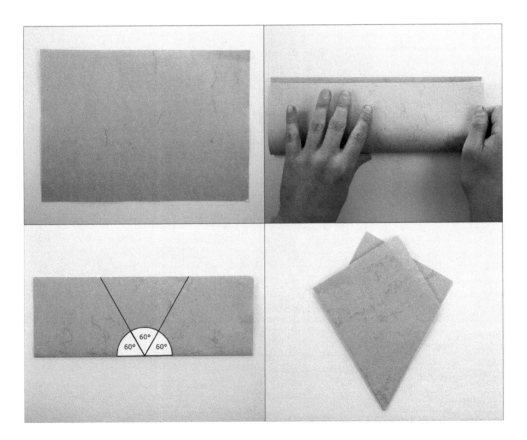

Figure 3.7.1 Forming a 60° intersection.

To free-fold an equilateral triangle twist, go through the same steps as you do to make a hexagon. You can use the ¼ marks or not, but it is good practice to try to eyeball 60°.

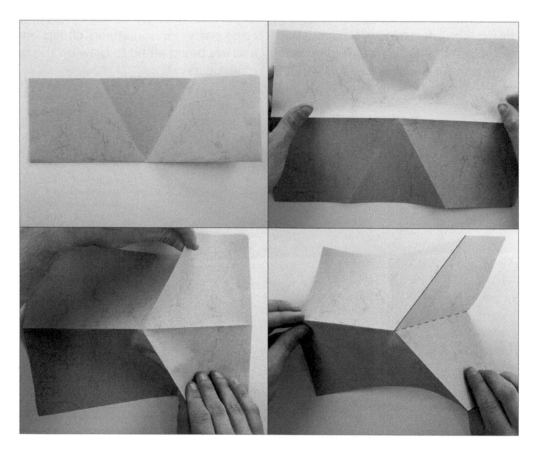

Figure 3.7.2 Inside-reverse fold.

Unfold entirely. You'll now employ a traditional origami maneuver called an *inside-reverse fold*. Turn the angled creases into mountain folds and the crease that bisects them at acute angles into a valley fold (Figure 3.7.2).

Figure 3.7.3 Folding the first pleat.

Bring the two facets on either side of the valley fold together to lie flat. Fold the top of the form down to create the first pleat, lining up the edges on the left to ensure that the resultant pleat is a parallel pleat. Lift the raw edge at the bottom to open and flatten the left side. The other pleats yet to be formed will rise off the table (Figure 3.7.3).

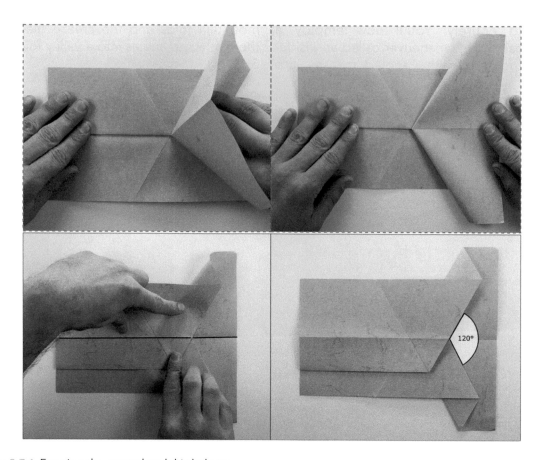

Figure 3.7.4 Forming the second and third pleats.

Open and begin to slightly flatten the children pleats. Do not crease flat; just get it into position. Flip the paper to the front and finish the flattening, lining up the creases marked in black in the third of Figure 3.7.4. This is a split at 120°.

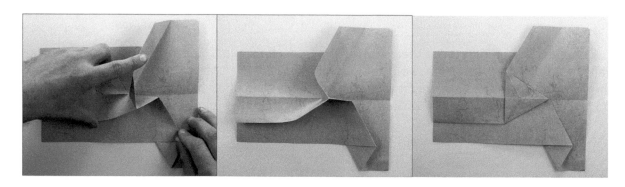

Figure 3.7.5 Reorienting the pleats into a triangle twist.

Reorient the pleat required to have them all oriented CCW to attain your equilateral triangle twist.

Unfolding and drawing on the lines, you can easily see the structure of the pleat intersection. The red marks are the valley folds with a CCW rotation. The pink mark are the valley folds with a CW rotation.

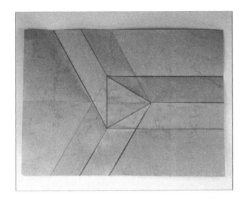

Figure 3.7.6 Pleat schematic for the triangle twist.

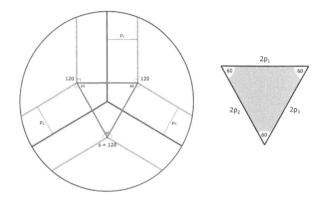

Figure 3.7.7 Equilateral triangle pleat pattern schematic.

You can say the pleat ends and the molecule starts when the pleat first hits another pleat. Connecting the borders of the pleats (valley and shadow folds), you get the orange triangle shown in Figure 3.7.7. The mountain folds are the perpendicular bisectors of the edges of that triangle, and during the collapse, the corners come together at a point in the finished form. Likewise, the two possible valley folds—in other words, the valley fold and the shadow of the pleats—come together and lie on top of each other in the finished form, regardless of which rotation you use.

I will call the angle of the split angle s. In the equilateral triangle twist, $s = 120°$. As the creation angle increases, the value for s decreases, and you end up with thinner children pleats. Likewise, as s increases, so does the width of the children pleats.

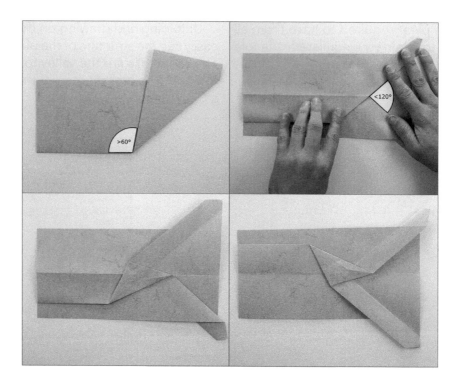

Figure 3.7.8 Free-folding an obtuse isosceles triangle twist. LEFT: CCW rotation. RIGHT: CW rotation.

As you change the value of angle s, the children pleat widths change relative to the parent pleat. The result is an isosceles triangle (Figure 3.7.8). If s < 120°, the isosceles triangle will be obtuse. If s > 120° the isosceles triangle will be acute.

Also, notice that unlike with the equilateral twist, the isosceles triangle twist platform changes depending on whether it has a CW or CCW rotation. The two triangle platforms for both the acute and obtuse triangles are shaded in light blue for CCW and aquamarine for CW.

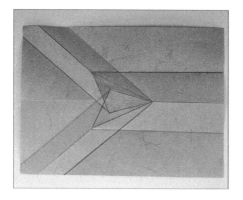

Figure 3.7.9 Obtuse isosceles triangle twist pleat schematic.

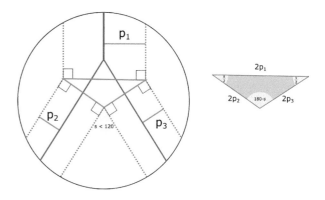

Figure 3.7.10 Acute isosceles triangle pleat pattern schematic.

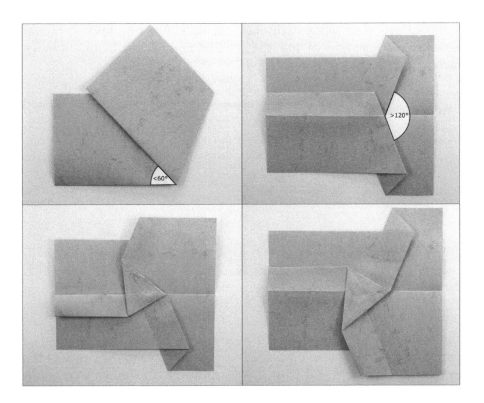

Figure 3.7.11 Free-folding an acute isosceles triangle twist. LEFT: CCW rotation. RIGHT: CW rotation.

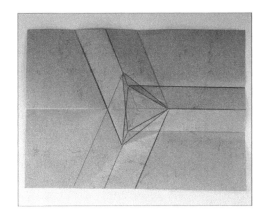

Figure 3.7.12 Acute isosceles triangle twist pleat schematic.

The diagrams, pleat schematics, and the orange triangle for the splits where $s < 120°$ and $s > 120°$ are shown in Figure 3.7.8–3.7.13.

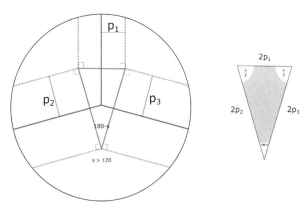

Figure 3.7.13 Obtuse isosceles triangle pleat pattern schematic.

This raises the question of what happens when the angles on either side of the split are not the same. This will result in a scalene triangle. The free-folding procedure is a little more complicated, but not much more so.

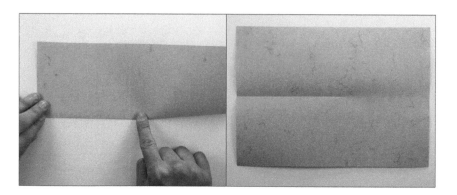

Figure 3.7.14 Free-folding the setup for the first pleat of a scalene triangle twist.

To create the scalene triangle twist, start by folding the paper in half, only creasing about halfway along the paper. Where you stop will be the intersection of the mountain folds of the pleats. Keeping the first pleat's mountain fold perpendicular to the edge of the paper allows you to use the edge as a guide for a parallel pleat (Figure 3.7.14).

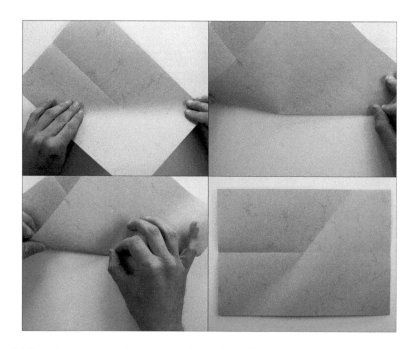

Figure 3.7.15 Free-folding the setup for the second pleat of a scalene triangle twist.

Now choose where you want the next mountain fold of the pleats to be. Fold that crease just far enough so that it intersects the first crease (Figure 3.7.15).

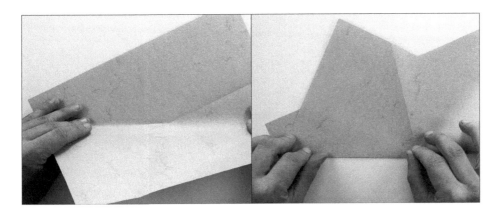

Figure 3.7.16 Free-folding the setup for the third pleat of a scalene triangle twist.

Then find where you want the mountain fold of the third pleat to be. Fold that pleat so that it intersects at the same point as the other two (Figure 3.7.16). Also, the angle between any two adjacent creases must not be greater than or equal to 180°.

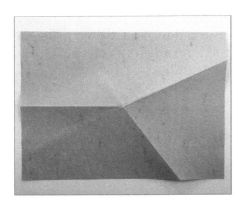

Figure 3.7.17 The three mountain folds of the twist.

Flip the paper back to the mountain fold side of the paper (Figure 3.7.17).

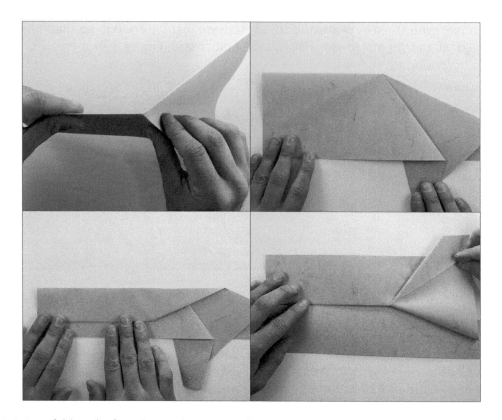

Figure 3.7.18 Free-folding the first pleat and setting up the second and third pleats.

Pinch the paper as shown in Figure 3.7.18 and lay it flat. It will naturally create this pleat intersection's analog to the inside-reverse fold.

Fold the top down to create the first pleat and open the form again as normally.

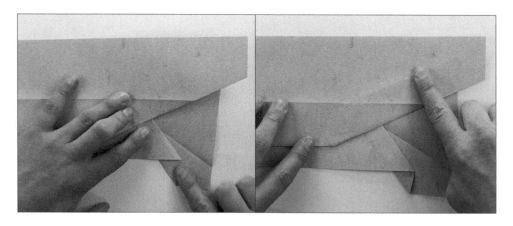

Figure 3.7.19 Free-folding the second and third pleats.

Flip the paper back over and set the pleats. Unfortunately, there are no guides to keep the children pleats parallel. With practice you will learn to eyeball parallel creases, or you can measure (Figure 3.7.19).

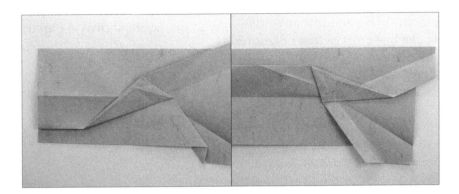

Figure 3.7.20 Finished scalene triangle twist. LEFT: CCW rotation. RIGHT: CW rotation.

When you finish the twists in either direction, these are the results CCW and CW.

Figure 3.7.21 Scalene triangle twist pleat schematic.

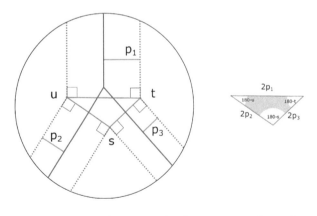

Figure 3.7.22 Scalene triangle pleat pattern schematic.

Figure 3.7.22 is a possible schematic result. We can name the other split angles u and t and get a more generalized orange shape in the middle. What happens when you have more creases intersecting at a point? Take some time to play with this concept, extending the process of folding the scalene triangle twist to have more mountain folds intersecting at a point. You will likely find that most intersections you attempt will not result in an FF molecule for one reason or another, but the pleats will at least lay flat. We will explore those reasons in subsequent sections.

257

3.8 The Normal Polygon

The orange triangle noted in the figures of Section 3.7 is a representation of what happens when the vertices of a pleat intersection come together to a point. The mountain folds—being perpendicular bisectors of that polygon's edge—bring the two corners of the orange polygon together, and the space outside of the valley fold and shadow represent the paper facets that are brought together. I will call this model the *normal polygon* of the pleat intersection, and it is a direct representation of that pleat intersection, assuming parallel pleats. The edges of the normal polygon can be taken from anywhere on the pleat, and are a direct cross-section of the pleats. It is called the normal polygon because its edges are normal or perpendicular to the creases of one of the parallel pleats.

Perhaps this begins to explain why the arrow twist has to use a double-width pleat. After all, when you put the pleat intersection of the arrow twist through the same process of connecting the vertices together, you end up with the figure shown in Figure 3.8.1.

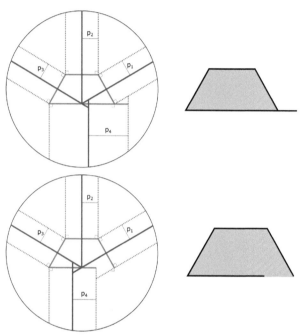

Figure 3.8.2 Schematic for forcing the pleats to be the incorrect proportion. TOP: One pleat too wide. BOTTOM: One pleat too thin.

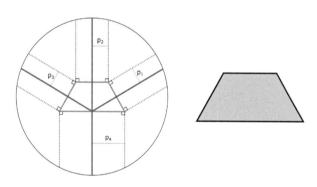

Figure 3.8.1 Schematic for the arrow twist.

What physically happens when you force the pleats to be thinner or wider than the seemingly desired lengths of the normal polygon?

Figure 3.8.3 Forcing the pleats to be the incorrect proportion. TOP: One pleat too wide. MIDDLE: One pleat too thin. BOTTOM: Every pleat the correct proportion (standing form of the arrow twist).

It results in an NFF pleat intersection of a type that I will call *warping*. Warping occurs when the cross-section lines of the pleats—from the shadow folds to the valley folds—do not create an enclosed polygon without an overlap (Figure 3.8.3).

The demonstration shows increasing or decreasing the width of p_4 without changing the widths of any other pleats. If you attempt to change the proportions of the pleat widths, the paper almost tries to correct itself, and it is impossible to keep the area around the molecule flat without it reverting to the "correct" pleat widths.

You might think that you can use this to work backward and create an FF pleat intersection using a normal polygon and creating pleats from that. To some extent you can, but there are other factors if you want the molecule to fold neatly. I present twenty polygon options as case studies for you to experiment with, of various levels of regularity and number of sides. I encourage you to draw or scan these pleat schematics to see how they fold. I also encourage you to draw your own normal polygons and create pleat intersections from that. You'll find that none of the following polygons create pleat intersections that suffer from warping issues, though there are other obstacles some will face.

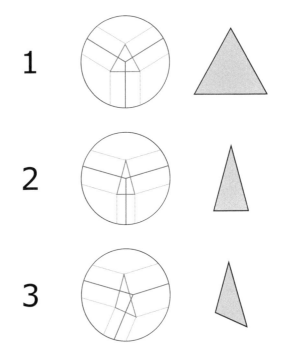

Figure 3.8.4 Triangular normal polygons case studies. TOP: Equilateral triangle. MIDDLE: Isosceles triangle. BOTTOM: Scalene triangle.

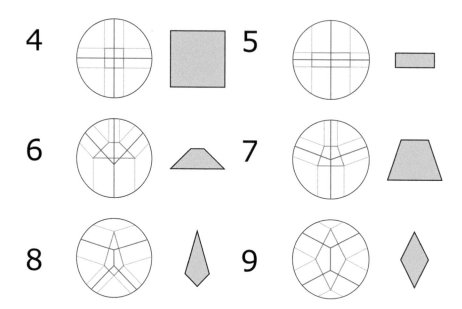

Figure 3.8.5 Quadrilateral normal polygons case studies. TOP ROW: Square and rectangle. MIDDLE ROW: Two different trapezoids. BOTTOM: Kite and rhombus.

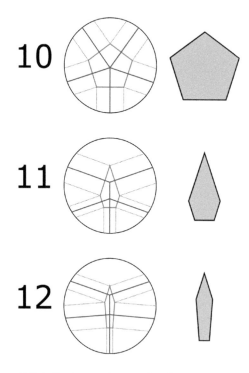

Figure 3.8.6 Pentagonal normal polygons case studies. TOP: Regular pentagon. MIDDLE: Stretched pentagon. BOTTOM: Alternate stretched pentagon.

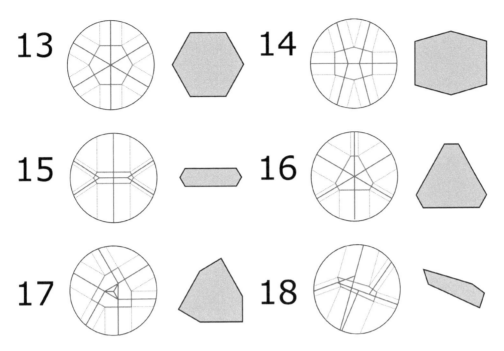

Figure 3.8.7 Hexagonal normal polygons case studies. TOP: Regular hexagon and stretched hexagon. MIDDLE: Alternate stretched hexagon and truncated triangle. BOTTOM: Skewed hexagon and irregular hexagon.

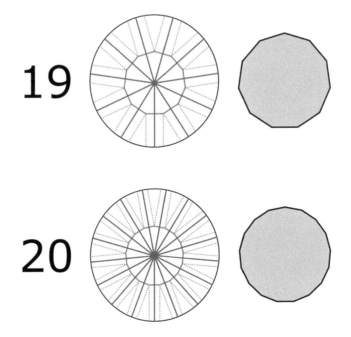

Figure 3.8.8 Higher-order normal polygons case studies. TOP: Regular hendecagon (11-sided). BOTTOM: Regular heptadecagon (17-sided).

Drifting and the Normal Model

The normal model can also include drifting in its analysis.

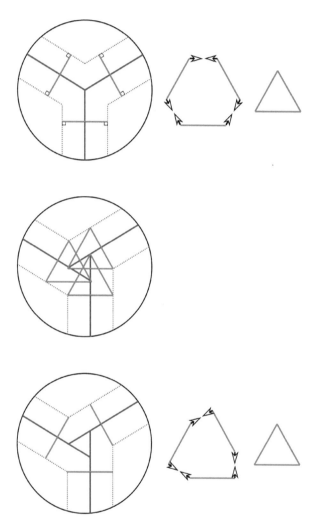

Now we'll look at the triangle spread twist and the pleats that go into it. This is because the edges of the normal polygon can be taken really from anywhere on the pleat. The top drawing of Figure 3.8.9 demonstrates this property.

The triangle spread has one pleat that is drifted relative to the others. If you draw the normal polygon, it actually results in three identical triangles as shown in the middle image of Figure 3.8.9. If you just have the relevant lines—the ones that are slid down until they hit vertices—the result is a disjointed polygon. Then, translating the edges together without rotation checks whether a polygon without holes or overlaps is created as shown in the bottom drawing of Figure 3.8.9.

3.9 The Circle Cutout Model

If you tried to fold the pleat schematic examples in Section 3.8 you'll have noticed that some of them become difficult to flatten when the mountain folds do not intersect at a single point. During the collapse of a twist, the valley fold and shadow lines are brought to lie atop one another. In addition to that, the vertices of the pleat intersection are drawn toward the intersection of the mountain folds. Let's see how that works with the arrow twist again.

Figure 3.8.9 The normal polygon with a drift. TOP: Regular normal polygon without a drift. MIDDLE: Disjointed normal polygon. BOTTOM: Bringing together the normal edges together to become a polygon.

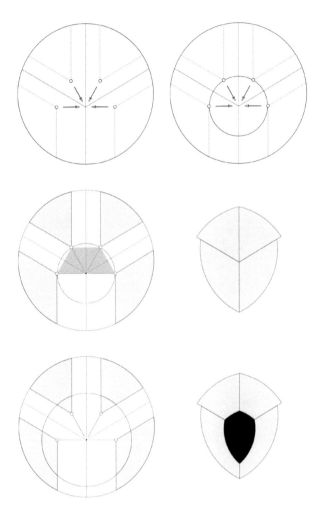

that point. Consequently, you can center a circle on the mountain fold intersections with a radius equal to the distance moved by the vertices (top row of Figure 3.9.1). This circle acts as part of another representation of the pleat intersection. Removing the molecule and the pleats and then pushing the grey facets (the part of the paper outside of the pleats) together, you get the arrangement shown in the right column middle row of Figure 3.9.1. This shows what the reverse side of the arrow twist looks like when it is folded, with the floors brought together. The molecule can be any size in relation to the pleats so long as it encompasses the vertices—the twist expansion is an example of a molecule enlarging relative to its standard pleat intersection. For this reason, you can enlarge the circle to encompass more of the paper and get a better picture of the shape that a molecule can take, shown in the bottom row of Figure 3.9.1. I will call this the *circle cutout model* of a pleat intersection.

Putting the normal polygon back in the pleat intersection—shown in the left column middle row of Figure 3.9.1—notice how the corners of the normal polygon line up perfectly with the vertices of the pleat interaction. This is by design; the vertices are equidistant from the intersection of the mountain folds, and the normal polygon is said to be *cyclic*, meaning that all of the vertices lie on a circle. The center of that circle is the same point as the intersection of the mountain folds and is a special point of the normal polygon. This point is called the *circumcenter*.

Figure 3.9.1 Schematics for the circle cutout. MIDDLE: With the smallest circle cutout. RIGHT: Slightly larger cutout and resultant convex curve.

The place where the mountain folds intersect is the only point where the vertices can move the same distance inward and connect at

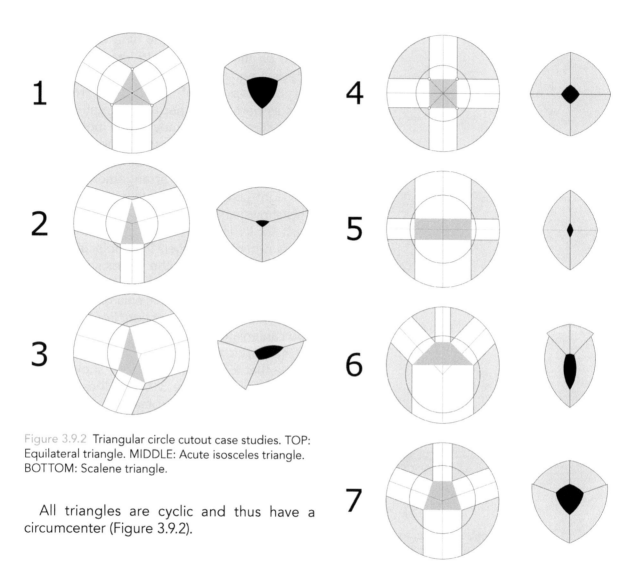

Figure 3.9.2 Triangular circle cutout case studies. TOP: Equilateral triangle. MIDDLE: Acute isosceles triangle. BOTTOM: Scalene triangle.

All triangles are cyclic and thus have a circumcenter (Figure 3.9.2).

Figure 3.9.3 Quadrilateral circle cutout case studies. TOP: Square. SECOND ROW: Rectangle. THIRD ROW: Trapezoid. BOTTOM: Shorter trapezoid.

All rectangles (and therefore, the square) are cyclic, as are all isosceles trapezoids. No parallelogram or kite other than a rectangle are cyclic, which is why the circle cutouts of the rhombi, parallelogram, and kite do not create a nice convex curve like those in Figure 3.9.3. We'll come back to those exceptions later in this section.

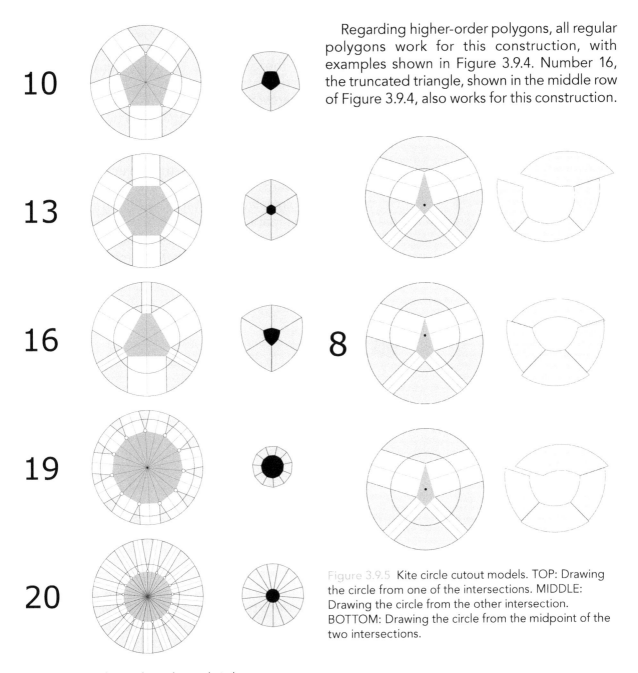

Regarding higher-order polygons, all regular polygons work for this construction, with examples shown in Figure 3.9.4. Number 16, the truncated triangle, shown in the middle row of Figure 3.9.4, also works for this construction.

Figure 3.9.4 Higher-order polygonal circle cutout case studies. TOP: Regular pentagon. SECOND ROW: Regular hexagon. THIRD ROW: Truncated triangle. FOURTH ROW: Hendecagon. BOTTOM ROW: Heptadecagon.

Figure 3.9.5 Kite circle cutout models. TOP: Drawing the circle from one of the intersections. MIDDLE: Drawing the circle from the other intersection. BOTTOM: Drawing the circle from the midpoint of the two intersections.

The ones that do not work all share the non-cyclic nature described previously.

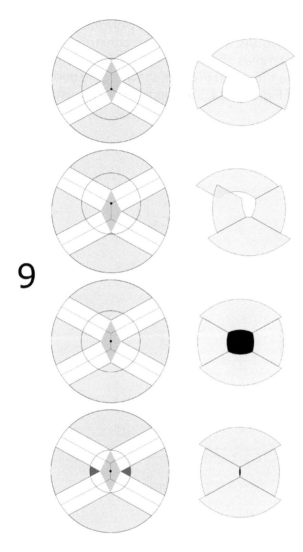

9

Figure 3.9.6 Rhombus normal polygons case studies. TOP: Drawing the circle from one of the intersections. SECOND ROW: Drawing the circle from the other intersection. THIRD ROW: Drawing the circle from the midpoint of the two intersections. FOURTH ROW: The smallest circle that can be drawn to encompass all vertices with clashing paper in red.

The rhombus, case study 9, has a curious condition where the floors are brought together. It doesn't fit together when the circle is drawn from one mountain fold intersection or the other, but when it is drawn from the midpoint of the segment created by those intersections, it seems to work, except for the fact that in folding, there are two triangles that clash during the collapse. These triangles are highlighted in red in the bottom drawing of Figure 3.9.6. When the vertices are brought inward, if the vertices are going to different points, layers of paper clash mid-collapse, and the paper cannot move any farther without displacing more paper. These intersections will lead to twists that are more complicated than otherwise; you will learn how to work around these complications in Section 3.11 with notches.

3.10 Molecule-to-Pleat Analysis

You now have an idea of what determines whether a pleat intersection will allow its pleats to flatten, and whether the molecule will intersect itself. But how does the molecule itself actually flatten into a twist? To begin to answer that, I will look at the molecule itself and work outward using MTP analysis. PTM and MTP analyses reveal different things about twist geometry and are both valuable in its understanding. Before I continue, let me define a special kind of twist, a *perfect twist*. Perfect twists are flat-folded twists where the only creases that go into its creation are the mountain and valley folds of its pleats and the outline of the polygon twist. From the six simple twists, the ones that are perfect are the triangle twist, the triangle spread, the hex twist, the hex spread, and the rhombic twist.

Figure 3.10.1 The six simple twists. BOTTOM RIGHT: Arrow twist, the only non-perfect of the six simple twists.

The arrow is different from the rest because it has a notch, extra paper beneath the twist platform. But why is that notch there in the first place? What differentiates the arrow twist from the others? For this, we will need a new model for a pleat intersection: the *desired perfect twist*, or DPT. To create a DPT of a pleat intersection, choose a rotation. Extend the valley folds of the

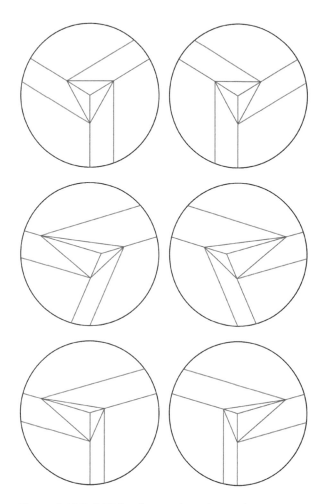

Figure 3.10.2 DPT for the equilateral triangle, and scalene triangle. LEFT: CCW rotation. RIGHT: CW rotation.

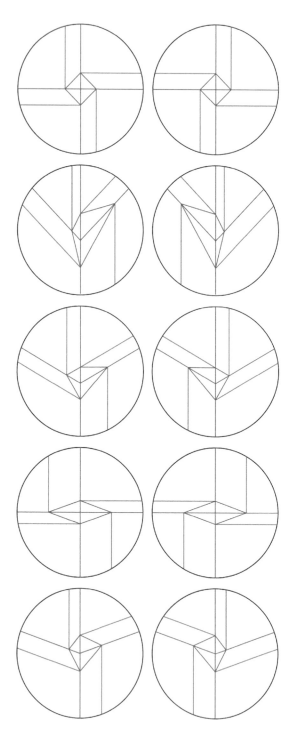

Figure 3.10.3 DPT for various cyclic quadrilaterals. LEFT: CCW rotation. RIGHT: CW rotation.

pleat schematic until they hit a mountain fold and remove the other rotation's valley folds. Where the valley fold of one pleat hits the next valley mountain fold of the next pleat in the rotation, it creates a vertex. Connect each of these vertices to form the DPT. The DPT represents the simplest possible molecule for that pleat intersection, assuming it can lie flat. Not all of them can, and you'll explore more of that in the next section. The DPTs for the cyclic polygon case studies follow in Figures 3.10.2–3.10.4.

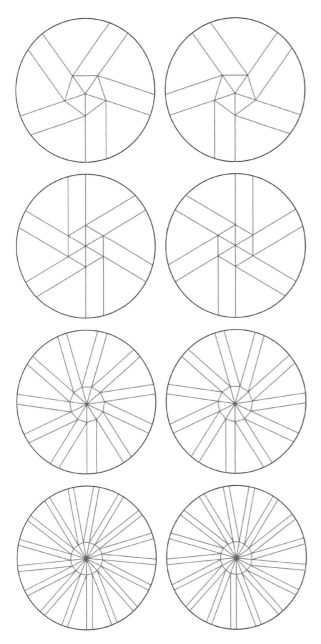

Not all of these CPs lead to flat-foldable twists, but if they do, that twist is a perfect twist. You will explore what the factors are for the DPT to create a perfect twist in Section 3.11.

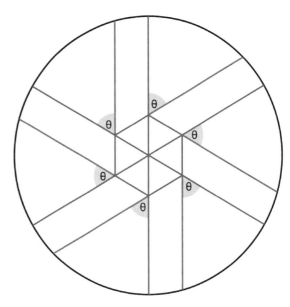

Figure 3.10.5 Exterior angles of a hexagon DPT being equal.

Several origami artists have documented that the outside angles of a perfect twist must be identical. Figure 3.10.5 shows the outside angle of a perfect twist being equal. Robert Lang offers an excellent explanation in *Twists, Tilings, and Tessellations* [1]. I will take a different—but mathematically equivalent—approach to understanding the angles in Section 3.12.

Figure 3.10.4 DPT for a regular pentagon, hexagon, hendecagon, and heptadecagon. LEFT: CCW rotation. RIGHT: CW rotation.

3.11 Sectioning Method of Twist Design

I will use the *sectioning method*—so called because it splits a polygon into sections—to derive the pleats that could create a given polygon twist. It involves starting with the polygon you want as the platform of the twist, defining the point you want the twist to rotate around, and then calculating the pleats that result in those metrics.

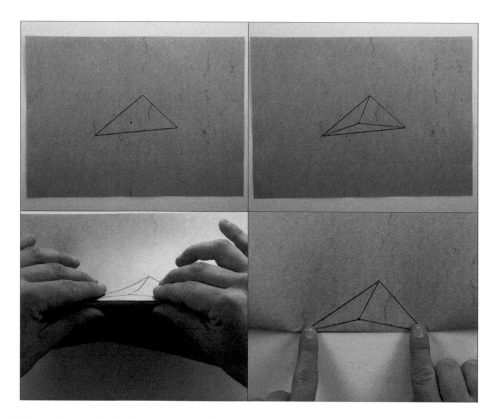

Figure 3.11.1 Sectioning method of perfect twist design. TOP LEFT: Scalene triangle and arbitrary point. RIGHT: Extending construction lines from that point to the vertices. BOTTOM ROW: Creasing each line as a mountain fold.

We'll start with a triangle, but note that this could be done with any convex polygon. Place an arbitrary point on the polygon and draw construction lines from that point to each of the corners of the polygon; this point will be called the *epicenter* of the twist and represents the point around which you want the polygon to twist. Pinch the edges of the polygon into mountain folds, one at a time, unfolding between each pinch. Don't extend the border creases further than the segment of the triangle edge (Figure 3.11.1).

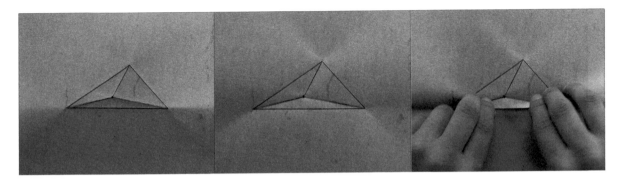

Figure 3.11.2 Backcreasing each of the lines. RIGHT: Pinching the edges of the triangle as mountain folds.

Do the same with the construction lines. Backcrease all six creases. Pinch one of the triangle borders as shown in the third photo of Figure 3.11.2.

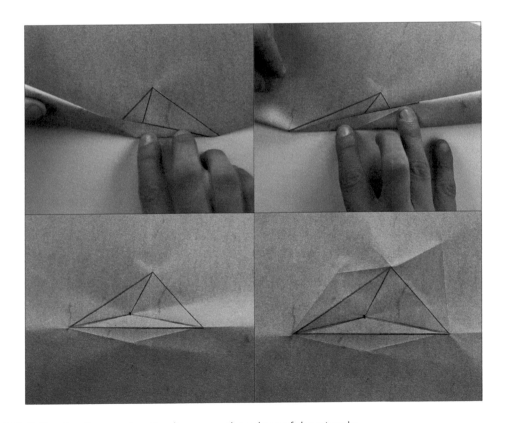

Figure 3.11.3 Reflecting the construction lines over the edges of the triangle.

271

Reflect each of the connecting construction lines over the edge of the polygon, unfolding between each. Pinch those reflections as valley folds. When you are finished with one side you will have something like the figure in the lower left photo of Figure 3.11.3. Do the same to the remaining two sides.

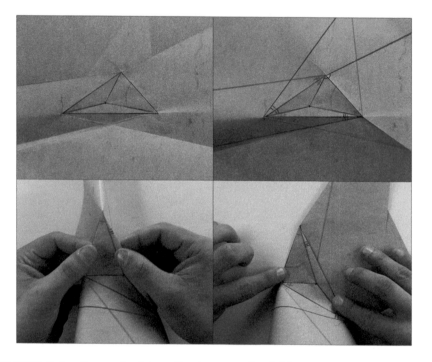

Figure 3.11.4 TOP ROW: Extending each of the reflections as valley folds. BOTTOM ROW: Finding the fourth crease by hand.

Extend these reflections to the edge of the paper. I drew the reflections to emphasize the creases in the second photo of Figure 3.11.4. The black lines will not be folded. The blue lines will be the mountain fold borders of the twist. The red lines will be the valley folds if you twist the pleat intersection CCW. The pink lines will be the valley folds if you twist the pleat intersection CW.

You are only missing the mountain folds of the pleats. Fortunately, it does not matter which rotation you use; the mountain folds of the pleats will be the same using this method. At each vertex, you have two mountain folds and only one valley fold for each rotation already constructed. This gives you enough information to figure out the mountain fold that will create an FF intersection at a given vertex. In order for an origami vertex to be FF, it must adhere to some basic conditions [1]:

- $M - V = \pm 2$; the number of mountain and valley folds in the vertex must differ by exactly two. We are working with a degree – four vertex – meaning there are four creases in it – so you are confined to either three mountain folds and a valley fold or three valley fold and a mountain fold. In this case, you'll make three mountains and a valley.
- The sum of every other angle in the intersection must add to 180°.

- The largest facet angle in the form must be surrounded by the same crease parity, either both mountain or both valley. A facet with this feature is called an *iso-facet*. If the crease parity is a mountain and a valley, it is called an *anto-facet*.

If you are given two mountain folds and a valley fold, you can figure out the third mountain fold with construction shown in Figure 3.11.5, the discovery of which is credited to origami artist Ilan Garibi [1].

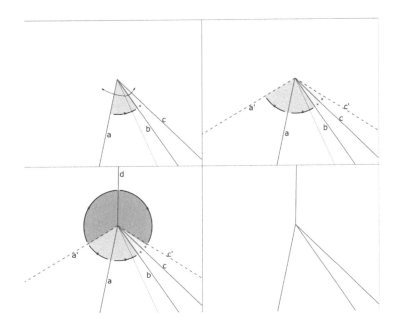

Figure 3.11.5 Geometric construction for finding the fourth crease. TOP ROW: Given two mountain creases *a* and *b* of a vertex and valley fold *c* of the same vertex, reflect *b* over *a* to find a' and *b* over *c* to find c'. BOTTOM LEFT: Bisect a' and c' to find the fourth crease in the vertex, crease *d*.

Now you have the ability to find the fourth crease in the vertex mathematically. You could have also done this by hand as the paper will naturally "calculate" the required location for the fourth crease.

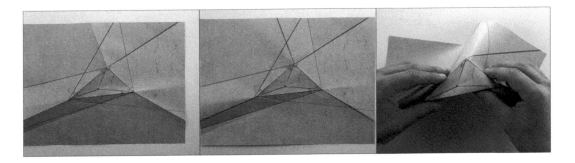

Figure 3.11.6 Forming the twist in one rotation or the other.

Do this procedure for each vertex, and you will now have every crease required to twist the form flat … mostly.

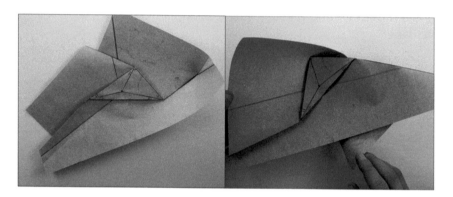

Figure 3.11.7 LEFT: CCW version of the twist. RIGHT: CW version of the twist.

The polygon twists around the epicenter, the point chosen at the beginning. Figure 3.11.7 shows the CCW and CW rotations of the example twist. The twist is FF to a point (extending outward), and then all of a sudden it stops being flat. This is because the sectioning method will often create nonparallel pleats. Any nonparallel pleat that extends from a twist either widens, thins, or remains the same width as it extends outward. I will call these *divergent pleats*, *convergent pleats*, and *parallel pleats*, respectively. The following figure shows the convergent and divergent pleats for a given rotation of the above twist (if convergent, the valley fold stops when it hits the mountain fold). The points of intersections are marked with circles and each pleat is colored orange if it is convergent or green if it is divergent.

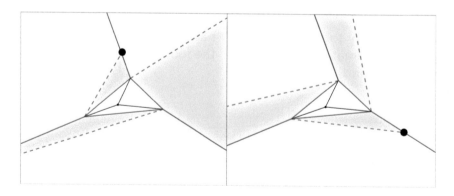

Figure 3.11.8 LEFT: CCW schematic. MIDDLE: CW schematic. RIGHT: CW rotation reverse side.

If any of the pleats are nonparallel, there must be at least one divergent and at least one convergent pleat. Any pleat intersection with a convergent pleat cannot be infinitely FF and will eventually lose its FF status if there is enough paper for the convergent pleat's mountain and valley folds to intersect. However, using the sectioning method the paper is FF until that point. I will call a pleat intersection where it is FF within a boundary that encompasses the molecule *regionally flat-foldable*, or *RFF*. If the paper edge is entirely within that boundary, the pleat intersection will be flat nonetheless.

You had many, many options for choosing your initial arbitrary point. Figures 3.11.9–3.11.11 show a few options folded, with the same triangle and a different initial point.

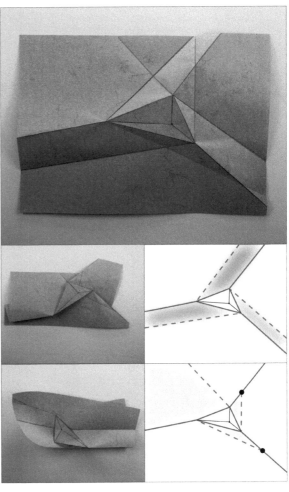

Figure 3.11.10 Alternate arbitrary point. TOP: Unfolded. MIDDLE: CCW rotation with all pleats being parallel. BOTTOM: CW rotation.

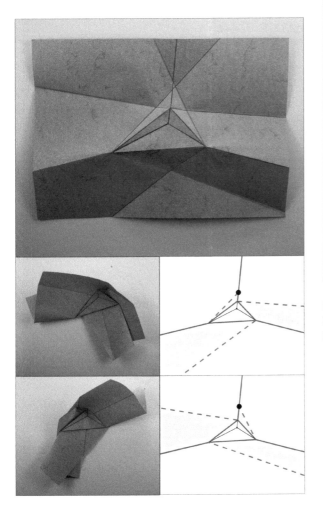

Figure 3.11.9 Alternate arbitrary point. TOP: Unfolded. MIDDLE: CCW rotation. BOTTOM: CW rotation.

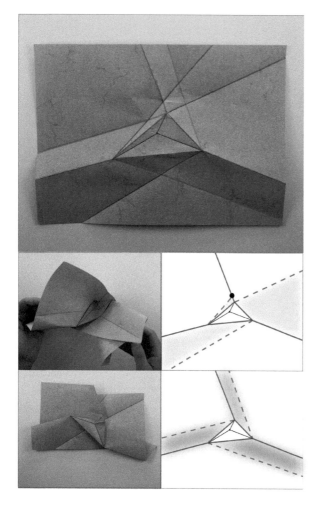

Figure 3.11.11 Alternate arbitrary point. TOP: Unfolded. MIDDLE: CCW rotation. BOTTOM: CW rotation with all pleats being parallel.

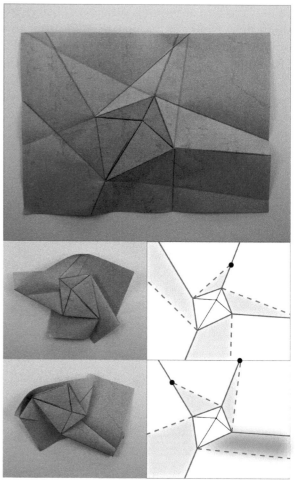

Figure 3.11.12 Sectioning method for an irregular quadrilateral. TOP: Unfolded. MIDDLE: CCW rotation. BOTTOM: CW rotation.

You can do the same with other convex polygons as well. Figures 3.11.12 and 3.11.13 show the sectioning method applied to an irregular quadrilateral and irregular pentagon, respectively.

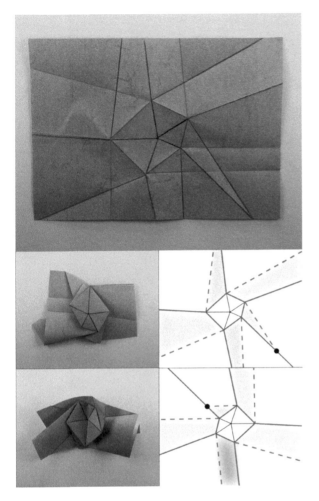

Figure 3.11.13 Sectioning method for an irregular pentagon. TOP: Unfolded. MIDDLE: CCW rotation. BOTTOM: CW rotation.

The Sectioning Method and Spreads

The epicenter does not need to be a point. You could have chosen a line or polygon, so long as you could interpret the line or polygon as a feature that occurs after the pivoting of the edges of the original polygon. Figure 3.11.14 shows examples of *epicenter polygons* and *epicenter line segments*. The first row of Figure 3.11.14 uses the same triangle as in Figures 3.11.8–3.11.11 with a *triangle epicenter* chosen to create the twist around. This results with a

negative space on the reverse of the paper in the form of the chosen triangle. You may recognize this as a similar reverse feature to the triangle spread and hex spread twists. If the reverse side of a twist has a polygonal negative space, I will say that it has a *polygonal spread*. More specifically, I will say the first example of Figure 3.11.14 has a *triangular spread*. The second example of the same figure shows an irregular quadrilateral with a *quadrilateral epicenter*, resulting in a *quadrilateral spread*. Likewise, the same can be done with an irregular pentagon, as shown in the third example of the same figure. The last two examples are of epicenter lines, resulting in *line spreads*. You may recognize this as a similar reverse feature to the rhombic twist.

Sectioning Method and Notches

It can be difficult to work with the sectioning method and arbitrary epicenters due to the tendency toward creating convergent pleats. If the pleat creases converge off the paper, you can still fold the twist flat, because an issue only arises at or beyond the point of convergence. However, if the creases of a pleat converge while still on the paper, the intersection is no longer FF from the point of convergence outward, and sometimes it is difficult to fold; the second-to-last row in Figure 3.11.14 is a notable example of a difficult-to-fold twist because it has two convergent pleats to wrestle into place.

There is a workaround to convergent pleats which I will call *notching*. This idea was mentioned briefly in Section 1.14 with the arrow twist. Applying the sectioning process to the rhombus platform and twisting about the epicenter marked in Figure 3.11.15, you end up with the schematic shown in the third drawing of the same figure. Notice that there is a quickly divergent pleat on the right—marked with green shading—and a quickly convergent pleat on the bottom—marked with orange

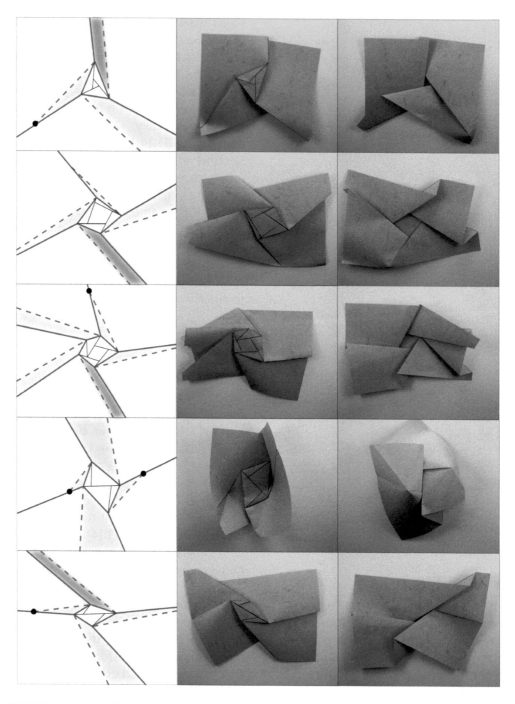

Figure 3.11.14 Sectioning method using epicenter polygons and lines instead of points LEFT COLUMN: CP. MIDDLE COLUMN: Folded front. RIGHT COLUMN: Folded reverse.

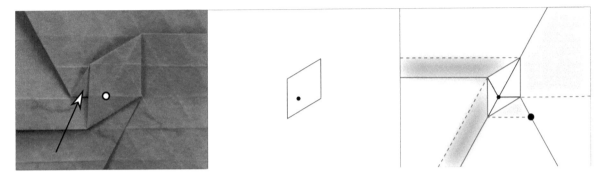

Figure 3.11.15 LEFT: The notch of an arrow twist. The rotation point is marked in white. MIDDLE: Starting polygon and epicenter for an arrow twist polygon. RIGHT: Resultant pleat schematic for the sectioning method.

shading—along with two parallel pleats on the left. This is very different from the pleat schematic for the arrow twist you saw in Section 1.14; that arrow twist had four parallel pleats! Notching is a way to change nonparallel pleats into parallel pleats and allows the sectioning method to result in an FF twist, so long as the epicenter chosen was a point—it does not work if the epicenter was a line or a polygon.

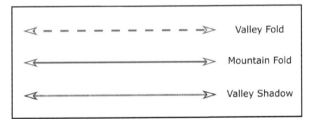

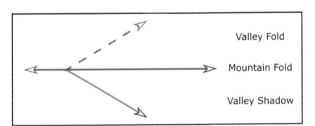

Figure 3.11.16 TOP: The valley shadow of a simple pleat. BOTTOM: The valley shadow of a mono, nonparallel pleat.

First, you have to identify the floors of the intersection, and this forces you to recall a term I touched upon in Section 1.7, the *valley shadow*, or just *shadow*. In parallel pleats, the distance between the mountain and valley folds defines the width of the physical pleat, but you have to account for the shadow—the valley that would exist if the pleat was oriented in the opposite direction—when figuring out the floors of a piece. For nonparallel pleats, the valley fold reflects over the mountain to find the shadow in the same way, as shown in Figure 3.11.16. The area of paper bounded by the shadow and the valley fold is the space that rises from the table during folding. Whatever paper lies outside of this bounded area remains on the table during the collapse and creates the floors.

279

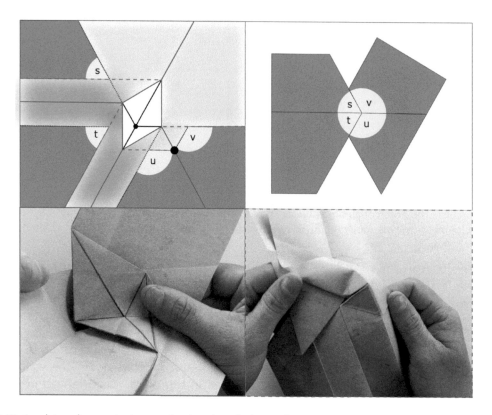

Figure 3.11.17 Applying the sectioning method to the platform of an arrow twist. BOTTOM ROW: The folded result of the sectioning process.

Figure 3.11.17 shows the pleat schematic for the arrow twist prior to notching. The bottom row of the same figure shows the folded version, and you can see the troublesome vertex resulting from the convergent pleat.

Even with the convergent pleats, the sectioning process makes the twist itself flat—it's specific parts of the paper around it that are NFF. Given this, it follows that the angles of the floors must sum to 360° since where they meet the paper is still flat. If you are rotating over a point (rather than a polygon or line spread), then you have the ability to rotate the floors around the epicenter—modifying the angle of the pleats to account for this movement—so long as you keep that sum relationship constant. In other words, you can change the angles of the nonparallel pleats until they are parallel. The easiest way to do this is to decrease the angle of the divergent pleat and increase the angle of the convergent pleat by the same amount. In the physical world, you can measure or eyeball parallel pleats and, with some practice, you can train yourself to approximate parallel pleats. With a computer program, you can be more precise, but the concept is similar in either case.

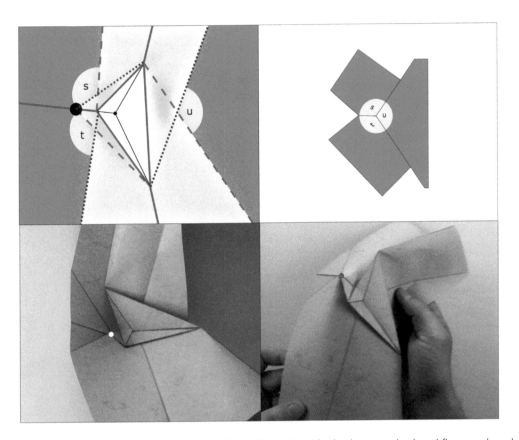

Figure 3.11.18 Scalene triangle twist. TOP ROW: Pleat schematic with shadows marked and floors colored. BOTTOM ROW: Inverting the convergent pleat's valley point to a mountain point.

Before delving into the arrow twist notching, I'd like to go over a simpler schematic, the CCW rotation of the scalene triangle from Figure 3.11.9. In this schematic, there is a quickly converging pleat, marked with a black circle in the schematic. In the bottom left photo of Figure 3.11.18, it is marked with an empty red circle, indicating a valley point; the corner is pointing away from you. The first thing to do is turn that valley point into a mountain point; make the corner point toward you instead. You will have to unfold the twist and refold to do this step. Once you invert that point—now marked blue to show a mountain point—find the line from the epicenter point through the point of convergence; in the image, this line is marked in blue.

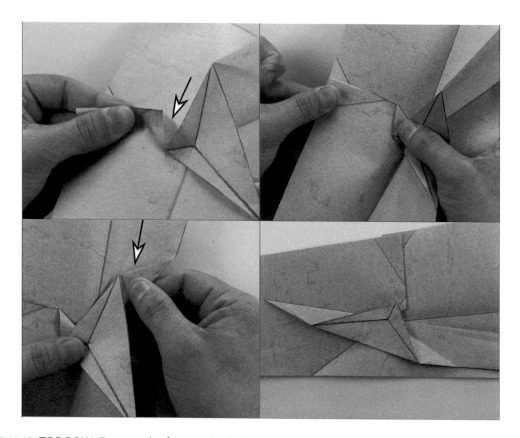

Figure 3.11.19 TOP ROW: Forming the first notched pleat. BOTTOM LEFT: Forming the second notched pleat. BOTTOM RIGHT: Finished notched twist.

Pinch that line to form the mountain fold of the notched pleat (Figure 3.11.19). Viewing the twist from the top, you can see the notch formed. This new "notched" mountain fold just needs a new valley fold. It can really be folded at any angle, but our goal is to turn it into a parallel pleat. Lay that pleat down flat as a parallel pleat, but do not lay any of the other pleats flat yet. You can eyeball this parallelism or measure as you desire.

If the sectioning method that created the twist only had a single convergent and divergent pleat, making the one convergent pleat parallel and keeping any other parallel pleats would automatically shift the divergent pleat into being parallel as well. This happens with our triangular case study, but if you have multiple convergent or divergent pleats, remember that you will have to notch each of the convergent pleats the same way before laying the entire twist flat.

Figure 3.11.20 Notching process for the arrow twist.

Now you have the tools to apply to the arrow twist. In the same way as the scalene triangle, turn the point of convergence from a valley point to a mountain point and pinch the line from the epicenter point through the point of convergence (Figure 3.11.20).

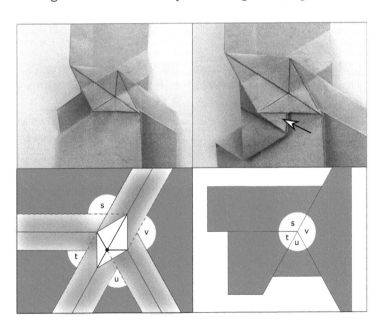

Figure 3.11.21 TOP ROW: Finished arrow twist—post notch. BOTTOM ROW: Pleat schematic for the arrow twist—post notch—with floors highlighted.

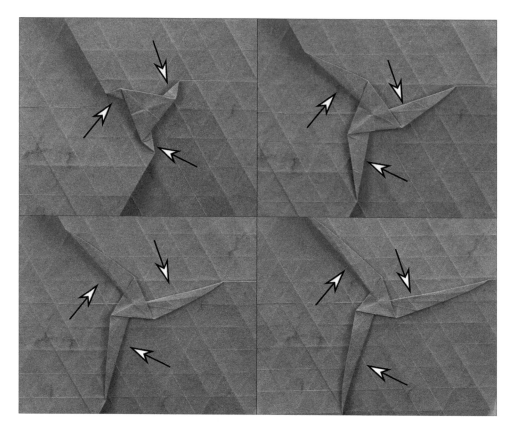

Figure 3.11.22 Backtwisting giving notches on each pleat by its nature.

When all of your pleats have been oriented correctly, the result is four parallel pleats, one of which has a notch. Viewing the pleat schematic, you see that the vertex where the floors meet still sums to 360°.

There are actually an infinite number of ways you can notch. In fact, you saw this when you learned about backtwisting in Section 2.11.

Backtwisting automatically imposes notches in twists that do not otherwise have them (Figure 3.11.22).

Notching also allows you to fold twists that otherwise wouldn't be feasible, such as those that have noncyclic normal polygons. Figure 3.11.23 shows some examples from the noncyclic case studies in Section 3.8.

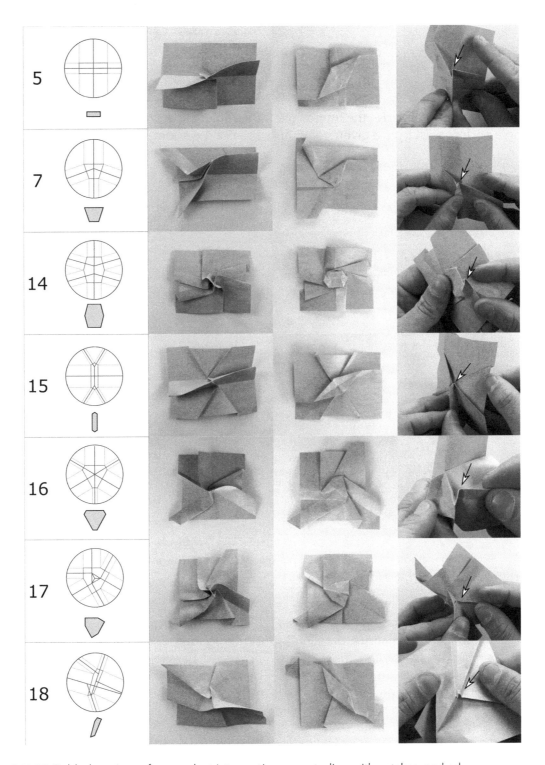

Figure 3.11.23 Folded versions of some pleat intersection case studies, with notches marked.

3.12 Brocard Points and the Shutter Skeleton of a Polygon

The sectioning method provides the framework for finding a perfect twist from a given polygon.

However, when chosen randomly, the arbitrary point can give nonparallel pleats, which makes the twist imperfect. The notch is a way to make the molecule and pleat intersection FF, but it's still not a perfect twist.

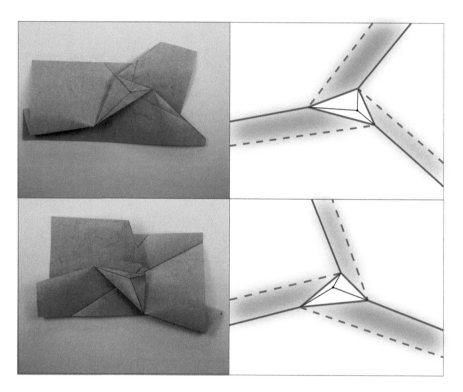

Figure 3.12.1 Sectioning method epicenter choice for a scalene triangle that results that have parallel pleats. TOP ROW: CCW rotation. BOTTOM ROW: CW rotation.

As you play with the sectioning method, you may stumble upon pleats that are parallel. In fact, in the six simple twists and most twists folded with a grid, the pleats are entirely parallel, without having to impose any notching. We saw this using the sectioning method shown in Figures 3.11.10 and 3.11.11.

Notice how the pleats in one rotation in either figure (but not the other rotation) ended up being parallel. So what makes the pleats become parallel? It has to do with the angle of reflection. For this, I will use a schematic for a polygon which I will call the *shutter skeleton*, due to its resemblance to a camera shutter closing.

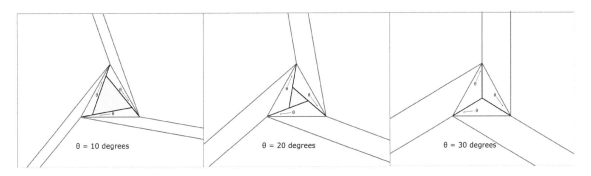

Figure 3.12.2 Shutter skeleton of an equilateral triangle. RIGHT: Critical point of the shutter skeleton with the critical angle of 30°.

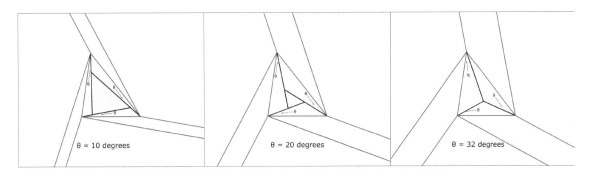

Figure 3.12.3 Shutter skeleton of a scalene triangle. RIGHT: Critical point of the shutter skeleton with the critical angle of 32°.

To create a shutter skeleton of a polygon, rotate the edges of the polygon inward in the same rotation (Figures 3.12.12 and 3.12.13). Each edge should rotate at the same angle (θ) and remain tethered to one of the vertices of the polygon. For a given polygon of n-sides, stop the rotation when the polygonal spread becomes a polygon of $<n$ sides. In the case of a triangle, the polygon will be another triangle that decreases in size as θ increases until eventually degrading to a point.

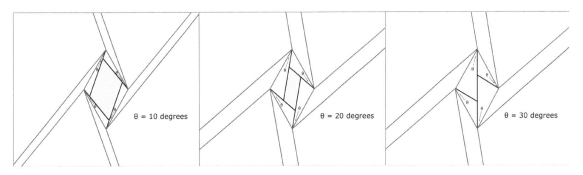

Figure 3.12.4 Shutter skeleton of a rhombus. RIGHT: Critical point of the shutter skeleton with the critical angle of 30°.

In the case of a nonsquare rhombus, this results in a line segment (Figure 3.12.4). You are unable to go any further, since the spread has decreased to $<n$ sides. Once the polygon degrades to $n-1$ or fewer sides, you cannot increase θ anymore and maintain an FF intersection; the bottom layers of the twist would clash. The angle where that occurs is called the *critical twist angle* [1].

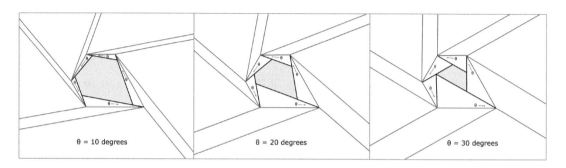

Figure 3.12.5 Shutter skeleton of a an irregular pentagon. RIGHT: Critical point of the shutter skeleton with the critical angle of 30°.

In the case of the irregular pentagon in Figure 3.12.5, the critical angle is 30°, and the result is the gem twist from the database, Section 3.4.

When the shutter skeleton construction degrades to a single point, it has a special name: the *Brocard point*. Every triangle has two

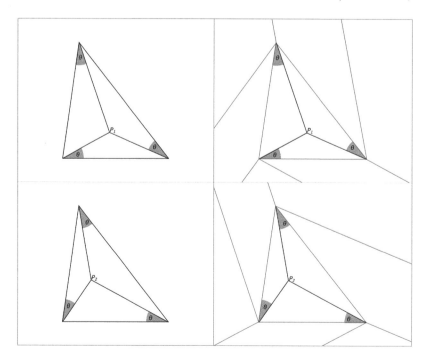

Figure 3.12.6 Definition of the Brocard points via construction. TOP ROW: First Brocard pont, labelled P_1. BOTTOM ROW: Second Brocard point, labelled P_2.

of them, called the first and second Brocard point, corresponding to either a CCW or CW rotation for its construction, respectively. The first Brocard point is found by pivoting each edge of the triangle CCW from the vertex it is connected to on the CW end until it reaches the critical angle, the angle where all of the pivoted edges intersect. The second Brocard point is found the same way, except the edges of the triangle pivot the opposite way and are tethered to the other vertex.

Applying the sectioning method to either Brocard point will result in a perfect twist—and thus, parallel pleats—and a rotation around that point so long as it is rotating in a corresponding manner to the Brocard point; that is to say, CCW rotation if the chosen point is the first Brocard point, and CW if the chosen point is the second Brocard point. If the first and second Brocard points are on the same point of the triangle, as is the case in an equilateral triangle, then either twist rotation will work. This is why when you reorient each of the pleats of an equilateral

triangle twist, you end up with an identical area of paper twisting. When you reorient a nonequilateral triangle twist, the result is a similar shape, but twists a different area of paper.

The proof by construction for why the same angle θ results in parallel pleats is shown in Figure 3.12.7. It is a local proof, which allows us to extend the proof to the other angle relationships in the polygon as well.

Another way to find the Brocard point of a triangle—my personal favorite—is to draw a circle through two adjacent vertices of the triangle, tangent to one of the edges, doing the same to each pair of adjacent vertices. Where those circles intersect is the Brocard point (Figure 2.12.8).

That works well for triangles. What about other shapes? Well, as you explore others, you'll find that they don't all have Brocard points. A polygon that has a Brocard point is called a *Brocard polygon*. Regular shapes—such as the square and equilateral hexagon—are Brocard polygons, as is each triangle. As the number of

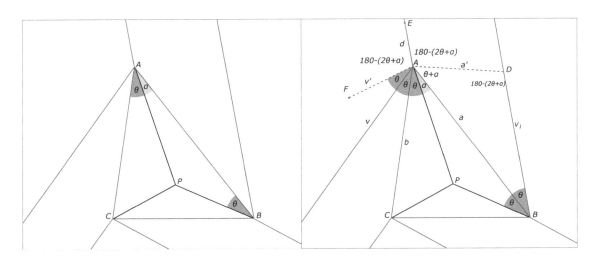

Figure 3.12.7 LEFT: Let the twist angle of the shutter skeleton of $\triangle ABC$ with starting point P be θ and angle $\triangle CAB - \theta$ be a. RIGHT: Fill in the angles of the construction lines using the procedure of the sectioning method. If $\triangle CAP = \triangle ABP$, then use $\triangle ADB$ to figure out $\triangle ADB$. Mountain fold d and valley fold $v_{1'}$ are parallel (and therefore the pleat is parallel) IFF $\triangle EAD = \triangle ADB$. This is the case when the twist angle of the shutter skeleton is consistent throughout the polygon.

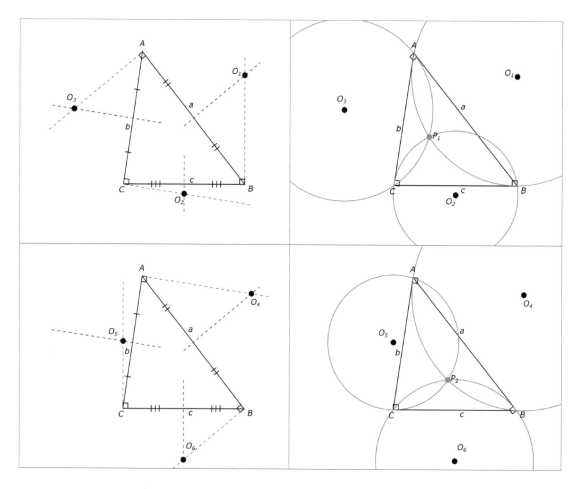

Figure 3.12.8 Construction of the Brocard point of △ ABC. TOP ROW: Construct the perpendicular bisector of segment *a* and a segment through point *B* perpendicular to line *c* and let their intersection be called O_1, Construct a circle centered on O_1 that passes through points *A* and *B*, which has segment *c* as a tangent. Repeat the process with the other two sides. The three circles will intersect at point P_1 the first Brocard point of the triangle. BOTTOM ROW: Replicate the same process as for the first Brocard point, but with the opposite rotation around the triangle to find P_2, the second Brocard point of the triangle.

sides of the polygon increases, the chances of you randomly constructing a Brocard polygon decrease rapidly.

If you apply the sectioning process of perfect twist design to a polygon, and if that polygon has a Brocard point, then choosing the Brocard point as the epicenter point results in a perfect twist that rotates around that point. To explore this, let's see some ways or finding quadrilaterals and higher-order Brocard polygons.

The first method uses constant side lengths. As you change the angles, eventually the construction circles—the ones running through two adjacent vertices, tangent to one of the sides—intersect at a point. In Figure 3.12.8 this happens when ∠DAB is roughly 106.91° [7].

Another method starts with the intersecting circles and constructs the triangle around it. You start with the two circles O_1 and O_2 which intersect at two points, *A* and *P*, defining *A* as a

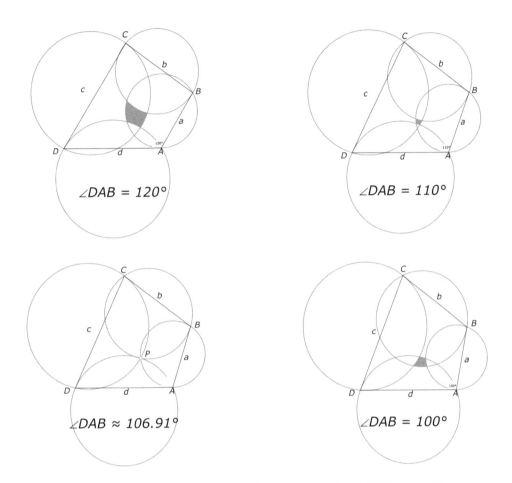

$\angle DAB = 120°$

$\angle DAB = 110°$

$\angle DAB \approx 106.91°$

$\angle DAB = 100°$

Figure 3.12.9 Construction of a Brocard quadrilateral with four given sides. TOP LEFT: ÐDAB = 120°. TOP RIGHT: ÐDAB = 110°. LOWER LEFT: ÐDAB = 106.91°. LOWER RIGHT: ÐDAB = 100°.

vertex of the polygon and P as the Brocard point of that polygon (Figure 2.12.10). In order for P to be the Brocard point, there has to a polygon side tangent to either O_1 or O_2 that is a chord of the circle not chosen; in the case of Figure 3.12.9, the side is tangent to O_1 and a chord of O_2, and creates point B, the second vertex of the polygon. Now, repeat the process with O_2, finding a tangent to point B that intersects with O_1, creating point C, the third point of the polygon. This is enough information to construct the third circle, O_3 using points C, P, and B. You can check your work by constructing the segment tangent to O_3 at point C as a chord of O_1. If that segment ends at point A, then P must be a Brocard point of that polygon.

The third circle is prescribed; you have no choice where it goes once you've drawn the first triangle segment. In fact, the only decisions you made were where the circles started. If they don't intersect or only intersect at a single point, the construction doesn't work. However,

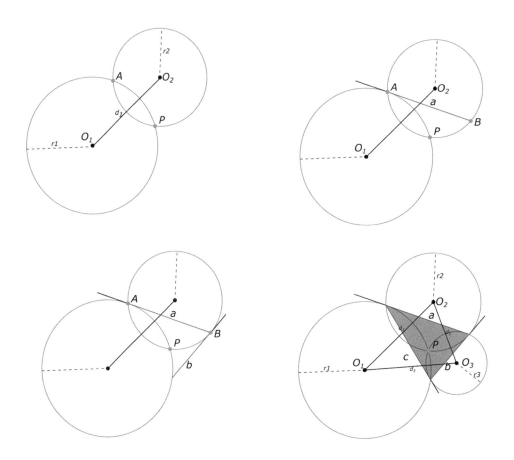

Figure 3.12.10 Construction of a triangle with a given side and its first Brocard point. TOP LEFT: Construct two circles, O_1 and O_2 that intersect at two points A and P. Label the distance between their origins d_1 and their respective radii r_1 and r_2. TOP RIGHT: draw segment a tangent to circle O_1 passing through A as a chord of circle O_2. Let the other end of the chord be point B. BOTTOM LEFT: Draw segment b tangent to Circle O_2 passing through B until it hits O_1 Label that point C. Construct circle O_3 to pass through points B, P, and C.

assuming they intersect at two points, the only independent variables were the radii of the circles r_1 and r_2 and the distance between their origins d_1. If we set $r_1 = 1$, then it's d_1/r_2 that determines the resultant polygon. But there is some range of ratios where the second line misses the O_1 entirely, and that makes the construction not work (Figure 3.12.11). You can get around that by constructing a third circle through two points, giving you an additional degree of freedom in its radius, r_3.

Given those choices, you can create line D, and if that hits the first circle, it will lead into creating a Brocard polygon that is a quadrilateral. If it misses, the smallest Brocard polygon it could give is a pentagon, and we repeat the process.

There is a great deal more to study with Brocard polygons, but these will give you some tools for understanding the Brocard points and which polygons allow for perfect twists around a point.

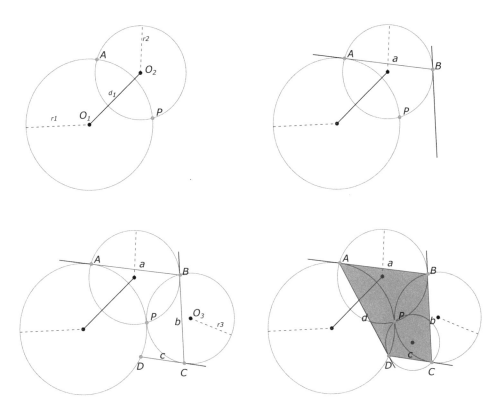

Figure 3.12.11 Construction of a quadrilateral with a given side and its first Brocard point. TOP ROW: Follow the same method as in Figure 3.12.10 but make the ratio d_1/r_2 be such that b does not intersect O_1. BOTTOM LEFT: Construct circle O_3 that passes through B and P. BOTTOM RIGHT: Following the same procedure as in Figure 3.12.10, if the tangent of O_3, segment c, hits circle O_1, finish the quadrilateral with first Brocard point P. If it does not, construct a fourth circle O_4, resulting in at least a polygon with $n = 5$ sides and continue until the polygon can enclose.

Chapter 4

Final Thoughts 2.0

This book has been an ongoing project of mine for six years, rooted in a time prior to that. During this time, I would sit at a small café in Aix-en-Provence, France with a stack of paper and a coffee and just fold. These café studies brought me to connect with the origami tessellation group on flickr, who introduced me to several lifelong puzzles. There's this mental itch that origami gives you, which never seems to go away, no matter how much you explore. The role of this book is (a) to bring others into the fold, (b) allow experienced tessellators to advance their own artwork, and (c) to promote this system to new audiences. I intend to release supplemental media to support this and carry on these tasks after the book's publication, because the work is ongoing. Even well-known tessellations are revealing their roles in new contexts, from textiles, to interior design and even modeling of galactic rotations [8]. There is exciting mathematics that is being written daily on the subject, and it is being recognized more and more as a medium for artistic expression. It is my hope that readers will take the techniques outlined here and develop their own patterns, puzzle out new solutions, and find enrichment in their creation. Keep folding, keep exploring!

Pleat Notation Thoughts

By Matthew Benet

There are many ways pleat behavior could be described. When creating our notation system, we explored several approaches, all numerical in nature. Why make a new notation system? Why not use an existing system, like crease patterns, to describe pleat behavior? Why use numbers at all? The reasons are numerous, but we'll list a few here. Numbers, like pleats, can be easily manipulated: we can use operations to easily show changes to a pleat; if there are patterns in the pleats, our numerical system can easily compress that information, resulting in a clearer description of the pleat pattern; changes to crease patterns are less obvious than changes to a tuple in our system, especially as we solidify operations; by using the tools and methods already established by mathematics to describe pleats, we access possibilities that are limited more by our imagination than by the math we're using.

Each representation choice has its benefits and shortcomings, and one isn't always better than another, but sometimes this new method can handle a particular problem more efficiently. I believe the creation of this notation is an important step in analyzing the behavior of pleats and their intersections.

Working with Ben to create an intuitive, mathematically-robust notation system has been a wild ride. It was important that our notation be open to expansion, yet comprehensive without additions—similar to how the equation defined by the Pythagorean theorem is useful to describe right triangles, yet can be expanded to describe any triangle in its general form, defined by the law of cosines. After countless hours of work—and several scrapped attempts—we settled on the notation used in this book. The current system is stronger than those that were abandoned, of course, but I believe this to be the case by orders of magnitude. All of our other systems fell apart under moderate scrutiny, but so far, this system has held up well. That is not to say our product is finished, only that we haven't found any glaring issues with it yet. All this in mind, our system as it stands has strong benefits.

Among these benefits are the operations it allows. We were careful to avoid using mathematically-established convention whenever possible. For instance, we have not yet formally defined exponentiation, addition, subtraction, multiplication, division, or any other mathematical operations, unary or binary. We do strongly believe, however, that several operations could be defined rigorously with little change to the system. For example, using relative notation, reorientation of a pleat could be notated with a negative sign. This could be considered inversion, which would fit with our intuition of how pleats are added together. Indeed, adding a pleat with its inverse would result in a mountain and valley separated by 0 distance, resulting in flat paper, what could be considered the identity pleat. Finally, associativity follows from the established rules of addition. All of this together appears to satisfy the four axioms of a group under addition.

There is one main reason that Ben and I haven't yet tackled the daunting task of rigorously defining operations in our notation system: We wanted to give the math and origami communities the opportunity to use the system and provide feedback on its strengths and weaknesses, as well as suggestions for the best way to utilize the full functionality of operations. With this input, we can ensure that operations in our system are intuitive and robust. We want our notation system to be as user-friendly and intuitive as possible, while still being rigorously defined.

Long-term, Ben and I want to finalize this system. It's hard to describe the excitement of having made so much progress, but there is still so much to do. The coming pages contain a more technical breakdown of the reasoning behind our notation system. We humbly ask that mathematicians and origami artists alike scrutinize this breakdown and submit their input to expand upon and improve our system.

We ask this in the hope that it may be helpful to the origami community as a whole, as well as mathematicians studying the field; our contact information can be found on Ben's website, www.brdparker.com. Thank you.

n-tuples with simple pleats:

Absolute form:

$$\begin{pmatrix} m_1 & m_2 & & m_2 \\ v_1 & v_2 & ,\dots, & v_n \end{pmatrix}$$

Relative form:

$$a\begin{pmatrix} m_1 & m_2 & & m_2 \\ v_1 & v_2 & ,\dots, & v_n \end{pmatrix}$$

Terminology

- We will refer to each full set of coordinates as an *n-tuple*, where *n* is the number of coordinates in our tuple. In this book, *n* is 6, the number of axes being used. We refer to them as *hextuples*.
- A *regular tuple* is one where the axes are evenly spaced out rotationally. Two key examples are the hextuples in this book, which use the regular hex grid, and 4-tuples (or quadruples) for a square grid. Unless direction of pleats is established, a tuple is assumed to be a regular tuple.
- Each element of the tuple is a set of pleats, called a *coordinate* of the tuple. In the case of a tuple of simple pleats, each set contains 1 element. Because each coordinate is a set of pleats, hextuples have 6 coordinates regardless of if each coordinate contains a simple pleat or a composite pleat.
- In regular tuples, when referring to a specific coordinate, we will describe them from their axis name, as we do with *x*-coordinates and *y*-coordinates for the Cartesian plane. For example, we would refer to the first coordinate in a regular tuple as the "$\mathcal{A}_1$-coordinate," the second as the "$\mathcal{A}_2$-coordinate," and so on.

Properties of the notation system:

- General benefits addressed or used in the book
  - As presented in the book, the notation is fully functional for any regular tuple without change. In these cases, the number of terms inside the parentheses is the number of rotational divisions. For example, our six-coordinate notation describes the hex grid, with $360°/6 = 60°$ between each pleat.
  - Can be used with composite pleats
  - The use of composite pleats paired with the ability to describe any regular grid means that any crease pattern can be described with a single tuple for any regular grid.
- Use of the null set
  - You may have noticed that we use the null set when there is not a pleat on a given axis. Use of the null set is valid because each coordinate is a set of pleats (most clearly seen when using composite pleats that lie parallel to a certain axis). Thus, each coordinate of a tuple could be considered a *set* of pleats, even when the set contains only one element, such as in the case of simple pleats. When there are no pleats for a given coordinate, we say the set of pleats for that coordinate is empty. Thus, we use the null set to denote an $\mathcal{A}$-coordinate whose set of pleats is empty.
- Further development/next steps
  - Angles of pleats
    - I strongly believe that this system can be expanded for pleats at *any* angle; in other words, it can apply to pleats that do not necessarily fit into a regular grid. There are multiple routes you can consider when denoting the angle of a pleat. Perhaps you could add a row

to each coordinate that has the degree measure. Perhaps there is an operation that could be performed that rotates the pleat, with the default measure of $0°$.
  - If you can change the direction of parallel pleats as noted previously, you should also be able to change the direction of a single mountain or valley. This would expand the notation to allow for non-parallel pleats.
  - If indeed the system can be expanded to describe parallel, non-parallel, simple, and composite pleats, there is no limit to the set of pleats and crease patterns that can be described using a single tuple in this notation.
- Operations
  - Exponentiation:
  - Perhaps exponentiation could be used as a shorthand in tuples for tessellations such as the 3^6 discussed in the book. This allows for more details than the traditional 3^6 notation such as the density of a tessellation.
  - Addition:
    - When used to describe pleat composition, as seems most intuitive: if addition is used to denote composition, there are two cases.
      1. The mountains are at different locations, which would simply lead to a composite pleat.
      2. The mountains have the same location, which would lead to a simple pleat where the valley locations are summed together to give a new pleat width and direction in relative notation.
  - When used to describe pleat drifting.
    1. In global notation, you would add the amount to be drifted to all

mountain and valley folds to be drifted.

2. In relative notation, you would only add the amount to be drifted to the mountain folds, as the valley fold location is described relative to the mountain fold.

– Making a tuple negative (placing a negative in front of it) could change all the valley fold locations from positive to negative (in relative form), which would reorient all the pleats into the opposite direction.

– This unary operation may follow all necessary conditions for turning a pleat into its additive inverse. We currently believe the most intuitive way to add pleats together (pleat composition) is to use the mountain fold value as the location and add the relative valley values together. Thus, a pleat composed with its additive inverse in this way would result in a valley at location 0 relative to the mountain. A mountain and valley on top of each other cancel each other out, resulting in the "identity" flat paper. This leads us to believe the group axioms of addition can be satisfied with our notation.

– Absolute value could be used to force all the valleys into the same direction, effectively reorienting all the pleats counterclockwise to turn any set of pleats into a twist.

– To rotate counterclockwise, we could simply take the negative absolute value of a tuple.

– Other operations

– I am confident that there are other operations that would be useful in our notation that we have yet to discover. I am also sure that there are some

benefits we were careful to build into the system that we've since forgotten about. The research is still new, and we've hardly touched upon much of the origami that this notation could describe.

• Benefits of the relative notation system:
  • a can easily be distributed (additively) into our relative tuple to give us our global tuple.
  • Reorientation is captured easily (simply multiply the v value by -1).
  • $a + m$ is the global location of the mountain fold.
  • v is the width and direction of the pleat.
  • This system is best for starting with a given mountain fold, and easily seeing the location of all other mountains and valleys relative to your starting point. This fits the intuition folders have when creating pleats.

• Benefits of the global notation system:
  • m is the global location of the mountain fold.
  • a can easily be factored (additively) out of our global tuple to give us our relative tuple.
  • v is the global location of the valley fold.
  • $v - m$ is equal to the width and direction of the pleat (this value is equal to the relative v value; if it is a negative number, the valley lies clockwise of the mountain, and if it's positive, it lies counterclockwise of the mountain).

• A hybrid system?
  • Pros:
    – Mountain values are global.
      – Mountains are used to describe the location of pleats. When adding pleats together, they only combine if the mountain values match; otherwise, they become a composite pleat.

- Valleys are relative to the mountains they are paired with.
  - This fits very well with our intuition of operations such as addition.
  - A hybrid system captures much of the benefit of each system.
  - It opens up coefficients for a different use in the future.
  - It eliminates the need for two systems.
- Cons:
  - It is underexplored.
  - Ben suggested something close to this, and I remember being quite opposed to mixing global and relative location. I don't remember why, but I think it was a good reason?

General n-Tuples of elements with m composite pleats in each element:

Absolute form:

$$\begin{pmatrix} m_{1.1}m_{1.2}\dots m_{1.m} & m_{2.1}m_{2.2}\dots m_{2.m} & \dots, & m_{n.1}m_{n.2}\cdots m_{n.m} \\ v_{1.1}\ v_{1.2}\ \dots\ v_{1.m} & v_{2.1}\ v_{2.2}\ \dots\ v_{2.m} & & v_{n.1}\ v_{n.2}\ \dots\ v_{n.m} \end{pmatrix}$$

Relative form:

$$a\begin{pmatrix} m_{1.1}m_{1.2}\dots m_{1.m} & m_{2.1}m_{2.2}\dots m_{2.m} & \dots, & m_{n.1}m_{n.2}\cdots m_{n.m} \\ v_{1.1}\ v_{1.2}\ \dots\ v_{1.m} & v_{2.1}\ v_{2.2}\ \dots\ v_{2.m} & & v_{n.1}\ v_{n.2}\ \dots\ v_{n.m} \end{pmatrix}$$

Glossary

Absolute Form
A form of notating pleats on a coordinate system that describes the location of each of the creases in the pleat – section 3.6, page 240

Anto Facet
A facet of a folded vertex that is surrounded by bounded by one valley and one mountain fold – section 3.11, page 273

Archetype Set
A designation on a hexagonal coordinate system for the format of the axes that contribute to a pleat intersection – section 3.3, page 209

Archetype Subset
A designation on a hexagonal coordinate system for the specific axes that contribute to a pleat intersection.
There are twenty-two archetype subsets that are flat-foldable – section 3.3, page 209

Archimedean Tiling
A tiling of regular polygons where every vertex is the same throughout the pattern – section 2.5, page 70

Arrow Twist
A simple twist that uses one double-wide pleat and three consecutive single-wide pleats, creates a rhombus platform; see Six Simple Twists – section 1.14, page 45

Axis
A set of parallel creases or lines – section 1.5, page 14

Backcreasing
The act of changing the parity of a crease – section 1.4, page 10

Backtwisting
Reversing the rotation of a twist while without reorienting the pleats – section 2.10, page 86 and section 2.11, page 91

Bar Unlock
The act of unlocking a lock and shaping it into a bar – section 2.2, page 59

Benet Notation
A method of documenting pleat intersections using six axis coordinates – section 1.16, page 49

Bilateral Symmetry
Shapes that can be folded in half, allowing the mapping of each half onto the other – section 2.5, page 69

Binary Operation
An operation that requires an initial state and a modifier with a value to create a resultant value – section 3.6, page 49

Bottom Pleat
In a pleat intersection, the side of a pleat that is on the opposite side from the molecule if the molecule is entirely on one side of the paper – section 2.32, page 176

Brocard Point
The point of a polygon where construction lines drawn from each of the vertices to that point results in the same angle for a given rotation. If that rotation is CCW, it is the first brocard point; if that rotation is CW, it is the second brocard point – section 3.12, page 286

Brocard Polygon
A polygon that has a brocard point – section 3.12, page 289

Button Molecule
A variant of the hex spread twist that modifies its standing form to create a button effect – section 2.10, page 86 and section 2.23, page 138

Children Pleats
In a split, the pleats that are created due to the split; see pleat splitting – section 1.7, page 31

Circle Cutout Model

A model of a pleat intersection that studies what the shape of the floors will be when the pleat intersection is folded – section 3.9, page 263

Circumcenter

In a cyclic polygon, the point of origin of the circumcircle for that polygon – section 3.9, page 263

Collapsing

The act of folding several previously-folded creases simultaneously – section 1.5, page 14

Composite Pleat

A pleat consisting of two or more pairs of mountain folds and valley folds – section 1.7, page 28

Convergent Pleat

A nonparallel pleat whose angle decreases as it goes away from a molecule – section 3.11, page 274

Crease Pattern (CP)

A drawing that shows every crease in a folded form – section 1.3, page 4

Critical Twist Angle

In a perfect twist, this is the maximum angle the twist angle will allow – section 3.12, page 288

Crooked Split

A split variant that uses a ridge line to create extra decoration – section 2.27, page 153

Cyclic Polygon

A polygon where one can draw a circle touching each of its vertices, called the circumcircle – section 3.9, page 263

Desired Perfect Twist (DPT)

Given a polygon, the crease pattern for a twist that would have that polygon as a platform, if one exists – section 3.10, page 267

Divergent Pleat

A nonparallel pleat whose angle decreases as it goes away from a molecule – section 3.11, page 274

Drifting

An action one can perform on a pleat that translates its position on the grid – section 1.7, page 30

Dual Tessellation

For a tessellation created with regular polygons, the tessellation that is created by connecting the incenters of those polygons – section 2.5, page 70

Epicenter

The point, line, or area around which a twist rotates – section 3.11, page 270

Equiangular

A polygon in which all of the angles are congruent – section 2.5, page 70

Equilateral

A polygon in which all of the side lengths are congruent – section 2.5, page 70

Exploding

In a tessellation, the process of expanding the space between the polygons at an even rate – section 2.6, page 74

Flat-Foldable (FF)

In a pleat intersection, a description that it can be folded flat – section 3.2, page 209

Flattened Pleat

An action of standing and separating the paper between the creases of a pleat – section 2.10, page 86 and section 2.13, page 95

Floor

The unfolded region on the paper between molecules and pleats – section 1.15, page 48

Free-Folding
Folding twists without a grid as a guide – section 3.7, page 247

Grid
A set of precreases that helps guide pleat intersections – section 1.2, page 3

Hex Twist
A simple twist that uses six pleats and creates a hexagon platform; see Six Simple Twists – section 1.11, page 37

Hex Spread Twist
A simple twist that uses six pleats with three drifted one grid line from the intersection of the other two and creates a hexagonal platform; see Six Simple Twists – section 1.12, page 42

Inside-Reverse Fold
A traditional fold that uses an angled precrease and backcreases the center crease to enclose paper inside of a vertex – section 3.7, page 249

Iso Facet
A facet of a folded vertex that is surrounded by bounded by either two mountain folds or two valley folds – section 3.11, page 273

Iso-Area Tessellation
An origami tessellation that is isometrically identical on both sides – section 2.10, page 86 and section 2.11, page 91

Kite Molecule
A variant of the rhombic twist with three pleats oriented in one rotation and one pleat oriented in the other rotation – section 2.10, page 86 and section 2.17, page 112

Leading Pleat
In a cluster of two or more consecutive axes, the pleat that is the most CCW of those axes – section 1.13, page 43

Line Spread
On the reverse of a twist, a negative space in the shape of a line segment – section 3.11, page 277

Molecule
A generic term for the displaced area of paper from a pleat intersection – section 1.8, page 33

Molecule Cluster
In an origami tessellation, several molecules surrounding a floor – section 2.3, page 63

Molecule-to-Pleat Analysis (MTP)
A study of pleat intersections where the molecule is given and the analyst must figure out the pleats that create it; see pleat-to-molecule analysis – section 3.2, page 209

Mono-Pleat
A pleat consisting of exactly one mountain and one valley fold – section 1.7, page 28

Non-Flat-Foldable (NFF)
In a pleat intersection, a description that it can not be folded flat – section 3.2, page 209

Nonparallel Pleat
A pleat where the mountain and valley folds are not all parallel to each other – section 1.7, page 28

Normal Polygon
A convex polygon where each edge is perpendicular to the creases in a pleat – section 3.8, page 258

Notch
In a twist that is not perfect, an extra layer of paper beneath the top platform – section 1.14, page 47

Notching
In the sectioning method, a workaround for an intersection with nonparallel pleats – section 3.11, page 277

Nub Molecule

A variant of the rhombic twist with two pleats oriented in one rotation and two pleats oriented in the other rotation – section 2.10, page 86 and section 2.17, page 112

Origami Tessellation

A tiling of molecules on a folded plane of paper or other folded material – section 2.7, page 74

Paper Memory

After creasing, the amount the paper remains folded after pressure is removed from the crease – section 1.4, page 12

Parallel Pleat

A pleat where all of the creases are parallel to each other – section 1.7, page 28

Parent Pleat

In a split, the pleats that was on the paper prior to and after the split process; see pleat splitting – section 1.7, page 31

Parity

A designation of whether a fold is mountain or valley, with regard to its viewed side – section 1.3, page 4

Partial Flattening

An act of flattening a pleat most of the way to a molecule but not fully; see flattened pleat – section 2.13, page 98

Perfect Twist

A molecule where the only creases are the mountains and valleys of the pleats that go into the form and the border of the twist – section 3.10, page 266

Periodic Tessellation

A tessellation where the tiles can extend indefinitely in the same arrangement across the plane – section 2.5, page 69

Pleat

An overlap in a plane or material – section 1.2, page 3

Pleat Intersection

An intersection of pleats that displaces an area of material – section 1.8, page 33

Pleat Intersection Composition

An operation that combines pleat intersections to create a single resultant pleat intersection – section 3.5, page 233

Pleat Unlocking

The action of crossing two pleats, locking the first one so it cannot be reoriented without unfolding the other – section 2.1, page 56

Pleat Orientation

The direction the mountain fold moves when the pleat is folded – section 1.7, page 29

Pleat Profile

A cross-section of a pleat – section 2.12, page 94

Pleat Pushing

The act of pinching several pleats by their mountain folds and pushing them together – section 2.36, page 189

Pleat Schematic

A method of documenting a pleat intersection that does not include the creases in the molecule – section 18, page 33

Pleat Splitting

An action one can perform on a pleat, creating two pleats from a single pleat, split at a point on the grid – section 1.7, page 31

Pleat Unlocking

The action of unfolding locked pleats – section 2.1, page 56

Pleat-to-Molecule Analysis (PTM)

A study of pleat intersections where the pleat intersection is given and the analyst must figure out the possible molecules it can create; see molecule-to-pleat analysis – section 3.2, page 209

Polygonal Spread
On the reverse of a twist, a negative space in the shape of a polygon – section 3.11, page 277

Precreasing
The act of folding and unfolding one or more creases with the intent on them to be refolded in a later step; see collapsing – section 1.5, page 14

Radial Symmetry
Shapes that can be folded in radially, allowing the mapping of one sector onto the other – section 2.5, page 69

Regionally-Flat-Foldable (RFF)
A pleat intersection that can be physically folded flat within a range, outside of which it cannot – section 3.11, page 2.75

Regular Polygon
A polygon that is equilateral and equiangular – section 2.5, page 70

Regular Tessellation
A tessellation where every polygon and vertex is congruent – section 2.5, page 70

Relative Form
A form of notating pleats on a coordinate system that describes the shape of the pleat separately from its location – section 3.6, page 240

Rhombic Twist
A simple twist that uses four pleats and creates a rhombus platform; see Six Simple Twists – section 1.13, page 42

Rhombille Tessellation
A tiling of rhombi where any given vertex is surrounded by six or three rhombi; the dual of the 3.6.3.6 tiling – section 2.5, page 72

Ridgeline
A sculpted path of a mountain fold to create decoration in tessellation – section 2.22, page 134

Sectioning Method
A method of creating specific twist platforms on demand – section 3.11, page 270

Self-Dual
A tiling that is a dual of itself; the square tiling is the only self-dual – section 2.5, page 72

Semiregular Tessellation
A tessellation using two or more types of regular polygon but the vertices are congruent throughout the plane – section 2.5, page 70

Shutter Skeleton
A schematic of a polygon where the edges pivot at the same angle and rotation inward, creating the effect of a camera shutter closing – section 3.12, page 286

Simple Molecule
A molecule composed of simple pleats – section 1.8, page 33

Simple Pleat
A pleat that is mono and parallel – section 1.7, page 28

Simple Flat Twist (Simple Twist)
A twist composed of simple pleats – section 1.8, page 33

Six Simple Twists
A curated set of twist that are helpful for novice folders to understand origami tessellations – section 1.8, page 33

Standing Form (of a molecule)
A form where the pleats of a molecule is folded, but the molecule is still three-dimensional – section 1.9, page 34

Standing Form (of a pleat)
A form of a pleat where the mountain fold is off the paper and each of the possible valley folds are folded at right angles to the plane – section 1.7, page 29

Sunk Twist
A twist where the platform is on the opposite side of the paper from the pleats – section 2.10, page 86 and section 2.18, page 114

Tessellation
A pattern of shapes that tile the plane without holes or overlaps – section 2.5, page 68

Tessellation Scaffolding
Folding a simpler tessellation in full before modifying it to create a new tessellation – section 2.14, page 101

Top Pleat
In a pleat intersection, the side of a pleat that is on the same side as the molecule if the molecule is entirely on one side of the paper – section 2.34, page 176

Trailing Pleat
In a cluster of several consecutive axes, the pleat that is the most CW of those axes – section 1.13, page 43

Triangle Grid (Hex Grid)
A grid type that results in congruent equilateral triangles throughout the paper – section 1.6, page 21

Triangle Twist
A simple twist that uses three pleats and creates a triangle platform; see Six Simple Twists – section 1.9, page 34

Triangle Spread Twist
A simple twist that uses three pleats with one drifted one grid line from the intersection of the other two and creates a triangle platform; see Six Simple Twists – section 1.10, page 36

Tulip Split
A split variant that transitions a double-wide pleat into three single-wide pleats – section 2.29, page 160

Twist
A type of molecule where there are platforms that rotate around an epicenter – section 1.8, page 33

Unary Operation
An operation that, when applied to an initial state only has one outcome on a resultant value – section 3.6, page 241

Valley Shadow
When a pleat is folded flat, this is the where the valley would be if the pleat were reoriented – section 1.7, page 29

Warping
A condition for non-flat-foldability where the normal polygon of a pleat intersection does not create a closed polygon – section 3.8, page 259

Photography Credit

Chris Bierlein (photographer)
Deconstruction Study – Front Cover

Alicia Gilbride (photographer)
Pages 107, 111, 126, 142, 137, 175

Uyen Nguyen (photo editor)
Various throughout

Christine Dalenta (photographer)
The Emergence – Page 195

Alessandro Beber (artist and photographer)
3.4.6.4, 4.3⁶.12, 4.6.12, 3.4.6.4 Rot., Menger 22, Menger 21, Penrose+, Perceptions, Promises, Space 1, Sqdod, Sqdod1, Sqdod2 – Pages 196–199

Thomas Petri (photographer)
Dod3, Dod34, Escape – Page 197

Robin Scholz (photographer)
Om – Page 198

Joel Cooper (artist and photographer)
Medallion – Pages 196-199

Ilan Garibi (artist and photographer)
Hidden – Page 200

Melina Hermsen (artist and photographer)
Bull Riding, Fraction, Eagle, Hidden, Falling, Lionheart – Page 201

Michał Kosmulski (artist and photographer)
Her Majesty's Tessellation, Star of David Tessellation, Star Interlace Tessellation, Union Jack – Page 202

Robert Lang (artist and photographer)
Flag, Hyperbolic Limit, Rock Climber – Page 203

Mike Marques (photographer)
Trashion Fashion Dress – Page 204

Christine Dalenta (artist and photographer)
Ridged Molecule Study 2, Snowflake Flagstone Molecule with Embedded Crumple Study – Page 205

Halina Rościszewska-Narloch (artist and photographer)
Eryngium Maritimum, Geastrum, Geastrum 1, UFO, Meadow – Pages 206, 207

Robin Scholz (artist and photographer)
All the Stars are Shining Tonight, Amizade – Page 208

Helena Verrill (artist and photographer)
Untitled – Page 208

Sarah Polucci (artist and photographer)
Origami – Back Cover

References

1. Robert J. Lang. *Twists, Tilings, and Tessellations: Mathematical Method for Geometric Origami.* CRC Press, Boca Raton, FL, 2018.
2. Eric Gjerde. *Origami Tessellations: Awe-Inspiring Geometric Designs.* A K Peters, Wellesley, MA, 2009.
3. Thomas Hull. *Project Origami: Activities for Exploring Mathematics, Second Edition.* CRC Press, Boca Raton, FL, 2018.
4. Alessandro Beber. *Origami New Worlds.* Self-Published, 2017.
5. Ilan Garibi. *Origami Tessellations for Everyone: Original Designs.* Self-Published, 2018.
6. Robert Lang. *Origami Design Secrets: Mathematical Methods for an Ancient Art, Second Edition.* CRC Press, Boca Raton, FL, 2012.
7. Roger Vogeler. Private conversation.
8. Mark Neyrinck. *Origami⁷: The Proceedings from the 7th International Meeting on Origami in Science, Mathematics, and Education; The Cosmic Spiderweb and General Origami Tessellation Design.* CRC Press, Boca Raton, FL.